A Pierre Costabel,
prêtre de l'Oratoire et historien des sciences,
aux encouragements et aux avis duquel ce livre doit beaucoup.

PRINCIPLES OF PHILOSOPHY

SYNTHESE HISTORICAL LIBRARY

TEXTS AND STUDIES IN THE HISTORY OF
LOGIC AND PHILOSOPHY

VOLUME 24

COLLECTION DES TRAVAUX
DE L'ACADÉMIE INTERNATIONALE D'HISTOIRE
DES SCIENCES
Nº 30

Franciscus à   Schooten Pr. Mat.

RENATVS DES-CARTES, DOMINVS DE PERRON. NATVS HAGÆ TVRONVM. ANNO. M.D.XCVI. VLTIMO DIE MARTII.

ad vivum delineavit   et fecit. Anno 1644.

Primus inaccessum qui per tot sæcula verum
Eruit è tetris longæ caliginis umbris,
Mysta sagax, Natura, tuus, sic cernitur Orbi
Cartesius. Voluit sacros in imagine vultus
Jungere victuræ artificis pia dextera famæ,
Omnia ut aspicerent quem sæcula nulla tacebunt.

CONSTANTINI HVGENII F.LY

RENÉ DESCARTES

PRINCIPLES OF PHILOSOPHY

Translated, with Explanatory Notes, by

VALENTINE RODGER MILLER and REESE P. MILLER

Huron College, University of Western Ontario, London, Ontario

D. REIDEL PUBLISHING COMPANY

DORDRECHT: HOLLAND/BOSTON: U.S.A.

LONDON: ENGLAND

Library of Congress Cataloging in Publication Data

Descartes, René, 1596–1650.
  Principles of philosophy.

    (Synthese historical library—texts and studies
in the history of logic and philosophy   ;   v.   24)
    Translation of : Principia philosophiae. 1644.
    With additional material from the French
translation of 1647.
    Includes index.
    1. Philosophy.   I. Miller, Valentine Rodger,
1939–   II. Miller, Reese P., 1934–   III. Title.
IV. Series.
B1863.E53M54   1983   100   82–18111
ISBN 90–277–1451–7

Published by D. Reidel Publishing Company,
P.O. Box 17, 3300 AA Dordrecht, Holland.

Sold and distributed in the U.S.A. and Canada
by Kluwer Boston Inc.
190 Old Derby Street, Hingham, MA 02043, U.S.A.

In all other countries, sold and distributed
by Kluwer Academic Publishers Group,
P.O. Box 322, 3300 AH Dordrecht, Holland.

D. Reidel Publishing Company is a member of the Kluwer Group.

Printed in The Netherlands

TABLE OF CONTENTS

TRANSLATORS' PREFACE

Descartes's *Principles of Philosophy* is his longest and most ambitious work; it is the only work in which he attempted to actually deduce scientific knowledge from Cartesian metaphysics, as he repeatedly claimed was possible. Whatever the success of this attempt, there can be no doubt that it was enormously influential. Cartesian celestial mechanics held sway for well over a century, and some of the best minds of that period, including Leibniz, Malebranche, Euler, and the Bernoullis, attempted to modify and quantify the Cartesian theory of vortices into an acceptable alternative to Newton's theory of universal gravitation. Thus, the *Principles* is not only of inherent and historical interest philosophically but is also a seminal document in the history of science and of 17th Century thought.

Principles of Philosophy was first published in Latin, in 1644. In 1647, a French translation, done by the Abbé Claude Picot and containing a great deal of additional material and a number of alterations in the original text, was published with Descartes's enthusiastic approval. Unlike some English translations of portions of the *Principles*, this translation uses the Latin text as its primary source; however, a good deal of additional material from Picot's translation has been included. There are several reasons for this. First, there is good evidence that Descartes himself was responsible for some of the additional material, including, of course, the Preface to the French translation. Second, the additional material sometimes provides an accurate and illuminating explanation of a point in the original Latin text. Finally, a number of modifications to the original, obviously made because Picot felt that Descartes had expressed himself incorrectly or too incautiously, are interesting in their own right since they supply some insight into the intellectual concerns and tensions of the times. Additional material from the French has, at times, been translated somewhat freely in order to enable it to fit smoothly into the Latin context without affecting the sense or syntax of the original Latin. In no case has material from the French edition been uncritically included in the text unless there is clear independent evidence that it reflects Descartes's own views, and all such inclusions in the text, except for punctuation, have been clearly indicated by being enclosed in braces.

Descartes hoped and intended that *Principles of Philosophy* would be used as a university textbook. The translators have the same hope for this edition. As a result, a number of annotations provide explanations of phrases or passages which would be obscure to one not well acquainted with the philosophical and scientific issues of the time or clarifications of passages which even an informed reader might find puzzling. Further, when it was felt that a philosophical or historical issue might be affected by the translation of a particular term or when, for a variety of reasons, it was not possible to find an exact equivalent in English, the Latin term has been noted and we have given alternative translations or an explanation. Other annotations supply explanatory quotations from Descartes's correspondence and other works or quotations which provide insight into Descartes's purposes and concerns. Although every effort has been made to make this translation as free from bias and interpretation as possible, certain passages defy intelligible translation without the imposition of an interpretation. Such passages have been noted and a very brief account of the grounds for the interpretation given. Finally, a few annotations result from the translators' reluctance to allow the considerable research time and effort expended on a particularly difficult passage to go unnoticed and our desire to spare the reader duplication of that effort.

Descartes's use of italics and upper-case letters has been retained whenever possible, and every effort has been made, by means of punctuation, to reproduce the logical structure of sentences which, in Latin, acquire that structure from the complexities of Latin grammar and syntax. The antecedent of a conditional statement, for example, is frequently separated from the consequent by a semicolon, even though the accepted rules of English punctuation would permit the use of a comma. Also, sentences which are extremely long and complex in the original are often rendered as a series of sentences. Within these limits, we have tried to render the original as literally and accurately as the restrictions of English usage will permit.

The primary source used for this translation is the revised Adam and Tannery edition of *Oeuvres de Descartes* (Paris: Vrin/C.N.R.S., 1966–1976). However, since neither that edition nor any other available edition is entirely free from error, a great deal of time and effort was spent comparing various editions and in research on the Latin and French usage of Descartes's time. This research, as well as enquiries into the works and sources which would have been available to Descartes, involved extensive use of the resources of the National Library of France, the French National

Archives and the Centre Alexandre Koyré, and was aided in part by research grants from the Centre National de la Recherche Scientifique (C.N.R.S.) and the Ministère des Affaires Etrangères of France. We would like to express our gratitude for that aid. Our thanks are also due to Mrs. G. B. McCall for her patience and care in preparing and correcting several versions of the typescript, to Anne Robiette who re-drew a number of the illustrations, and to the many friends and colleagues whose interest in and enthusiasm for this project sustained us over the years.

We would also like to thank Vrin/C.N.R.S. for permission to use material from their revised edition of Adam and Tannery. All references in the text to Adam and Tannery are to that edition, and notations which refer to Articles or notes by number alone indicate an Article or note in that Part of the *Principles* in which the notation itself occurs. Material enclosed in square brackets does not occur in the original and has been added by the translators; material in parentheses does appear in the original, although it may not have originally been parenthesized.

Finally, no translation of this length and complexity can claim to be entirely free from error, and this work is doubtless no exception. There may well be cases in which a misprint in the Latin or French text has gone unnoticed, places where we have mistakenly attributed material to the Latin or to the French text, or instances in which a term or phrase has simply been mistranslated. Scholars noting such errors are asked to be kind enough to communicate them to the translators.

Huron College,
University of Western Ontario,
London, Ontario,
Canada N6G 1H3

VALENTINE RODGER MILLER and
REESE P. MILLER

TO THE MOST SERENE PRINCESS ELIZABETH, ELDEST DAUGHTER OF FREDERICK, KING OF BOHEMIA, COUNT PALATINE, AND ELECTOR OF THE HOLY ROMAN EMPIRE

Most serene Princess,

I have perceived that the greatest reward of the writings which I have so far published is that you have deigned to read through them, and that on account of them I have been admitted to your acquaintance, and have learned that your qualities are such that I think it to be in the interest of the human race for them to be set forth as an example for posterity. It would not be appropriate for me either to flatter, or to affirm anything which I had not sufficiently investigated; especially here, where I am about to attempt to lay the foundations of truth. And I know that the unaffected and simple judgment of a Philosopher will be more pleasing to your noble modesty than the more elaborate praises of more ingratiating men. Accordingly, I shall write only those things which I know to be true either from reason or experience; and I shall philosophize here in the dedication in the same way as in all the rest of the book.

There is a great distinction between the true and the apparent virtues; and moreover, among the true virtues, between those which result from a meticulous comprehension of things and those which are associated with some ignorance. By apparent virtues, I understand certain rather uncommon vices, which are contrary to other more familiar vices; and which, being further from the latter than are the intermediate virtues, are therefore accustomed to be more greatly praised. Thus, because there are many more who timidly flee dangers than there are who rashly throw themselves into them; temerity is opposed to the vice of timidity as if it were a virtue, and is commonly esteemed more highly than true courage. Similarly, the prodigal are often more highly valued than the liberal; and none more easily acquire a great reputation for piety than the superstitious or the hypocritical.

Moreover, many true virtues result not solely from the comprehension of what is right, but also from some error: thus, goodness often arises from simple-mindedness, piety from fear, and courage from desperation. And these true virtues are distinct from one another, and also are designated by distinct names; but those pure and unmixed ones which spring solely from the comprehension of what is right all have one and the same nature, and are included under the name of wisdom. For whoever has the firm and effective will to always use his reason correctly, as far as is in his power, and to pursue all that which he knows to be best, is truly wise as far as his nature permits. And through that single trait, he has justice, courage, moderation, and all the other virtues united together in such a way that none stand out above the others. And although these virtues may accordingly be much more excellent than those which are made to stand out by some admixture of vices; they are not however usually extolled by as much praise, because they are not so well known to the multitude.

Besides, while two things are required for the wisdom thus described, i.e., perceptiveness of the intellect and inclination of the will: no one is incapable of that which depends upon the will, but some people have a much keener intellect than others. And although it ought to suffice, for those who are by nature slightly slower, that (even though they may be ignorant of many things) they can nevertheless be wise and thereby very pleasing to God provided that they retain a firm and constant will to omit nothing by means of which they may reach the comprehension of what is right, and to pursue all that which they judge to be right within their limitations: however, those in whom is found the keenest intellect and the greatest zeal for knowing the truth along with the firmest will to act rightly are much more outstanding.

And that this highest zeal is indeed in your Highness is obvious from the fact that neither the distractions of the court nor the customary upbringing which usually condemns girls to ignorance could prevent you from discovering all the liberal arts and all the sciences. Then, the supreme, and indeed incomparable, perspicacity of your mind is apparent from the fact that you have thoroughly probed all the mysteries of these sciences, and have mastered them in the shortest time. And I personally have a still greater proof of this, since I have so far found that only you understand perfectly all the treatises which I have published up to this time. For to most others, even to the most gifted and learned, my works seem very obscure; and with almost all it happens that if they are versed in Metaphysics, they shy away from Geometry; whereas if they have studied Geometry, they do not grasp what I have written about First Philosophy: I know of no mind but yours to which all things are equally evident, and which I therefore deservedly term incomparable. And when I consider that such a varied and perfect comprehension of all things is not in some already aged Sage,[1] who has employed many years in study, but in a Royal lady, who in beauty and youth resembles Charis rather than grey-eyed Minerva or any of the Muses; I cannot help being seized with the highest admiration.

Finally, I notice that nothing is required for perfect and sublime wisdom, either with regard to comprehension or to will, which does not shine forth in your conduct. For there appears in it, along with majesty, a certain exceptional benevolence and gentleness, which is assaulted by the perpetual injustices of fortune but never provoked or daunted. And this has conquered me to such an extent that I not only think that this Philosophy of mine ought to be dedicated and consecrated to the Wisdom which I perceive in you (because my Philosophy itself is nothing other than the study of wisdom); but that I also have no greater wish to hear myself called a Philosopher than to be called

> The most devoted admirer of
> your Most Serene Highness
> DES-CARTES

[1] Literally: "Gymnosophist".

LETTER FROM THE AUTHOR
TO THE TRANSLATOR OF THIS BOOK

(which can serve here as a Preface)

Sir,

The translation of my Principles which you have taken the trouble to prepare is so clear and so thorough as to give me hope that my work will find more readers in French than in Latin, and be better understood. My only fear is that the title may discourage many who have not been nurtured in the humanities, or else who have a poor opinion of Philosophy because that which they were taught did not satisfy them. Therefore, I believe that it would be advisable to add a Preface here, which would announce the subject of the Book, the intention which I had in writing it, and the benefit which can be derived from it. However, although I ought to write this Preface because I must know these things better than anyone else; I cannot persuade myself to do more than to summarize here the principal points which it seems to me ought to be treated in it. And I leave it to your discretion to present to the public as much of my summary as you judge appropriate.

I should have liked to begin by explaining what Philosophy is, beginning with the most basic points, for instance: that the word 'Philosophy' means 'the study of Wisdom', and by 'Wisdom', we understand not only prudence in our affairs, but also a perfect knowledge of all the things which man can know for the conduct of his life, the preservation of his health, and the discovery of all the arts. And for this knowledge to be thus perfected, it must necessarily be deduced from first causes; so that, to study for its acquisition (which [study] is properly called "philosophizing"), one must begin by searching for these first causes, that is, for Principles. And these Principles must meet two conditions: first, they must be so clear and so evident that the human mind cannot doubt of their truth when it attentively considers them; and second, the knowledge of other things must depend upon these Principles in such a way that they may be known without the other things, but not *vice versa*. And then, one must attempt to deduce from these Principles the knowledge of the things which depend upon them, in such a way that there is nothing in the whole sequence of deductions which one makes from them which is not very manifest. In truth, only God is perfectly Wise, that is to say, has a complete knowledge of the truth of all things; but it can be said that men have more or less Wisdom in proportion to the amount of knowledge which they have of the most important truths. And I think that there is nothing in these remarks with which all learned men do not agree.

I would subsequently have invited consideration of the usefulness of this Philosophy, and would have shown that, since it extends to everything which the human mind can know, we must believe that it alone distinguishes us from the most savage and barbaric peoples, and that each nation is the more civilized and cultured the better men philosophize there; and that, consequently, the greatest possible good for a State is to have true Philosophers. And furthermore, it is not merely useful for each individual man to live among those who apply themselves to this study, but incomparably better to apply himself to it; just as it is undoubtedly far better to use one's own eyes to guide oneself (and by the same means to enjoy the beauty of colors and light) than to keep one's eyes closed and follow another's guidance; though even the latter is better than keeping them closed and having only oneself for a guide. To live without philosophizing is, properly speaking, to have one's eyes closed and never attempt to open them; and the pleasure of seeing all the things which our sight reveals cannot be compared to the satisfaction given by the knowledge of those things which one discovers through Philosophy. And finally, this study is more necessary to regulate our morals and to guide us in this life than is the use of our eyes to guide our steps. Brute beasts, which have only their body to preserve, constantly busy themselves in seeking nourishment for it; but men, whose principal part is the mind, ought to give their principal care to the search for Wisdom, which is the mind's true nourishment. And I am sure as well that there are many who would not fail to do so, if they had hope of succeeding, and if they knew the extent of their capability. There is no soul with even a trace of nobility which remains so strongly attached to the objects of the senses that it does not sometimes turn aside from them to wish for some other greater good; even though it often does not know in what this greater good consists. Those whom fortune favors most, who have an abundance of good health, honors, and riches, are no more exempt from this desire than the rest; on the contrary, I am convinced that it is they who yearn the most ardently for another good, higher than all they possess. Now this supreme good, considered by means of the natural reason without the light of faith, is nothing other than the knowledge of the truth through its first causes, that is to say, Wisdom, of which Philosophy is the study. And, because all these things are entirely true, it would not be difficult to convince men of them, if they were correctly demonstrated.

But because we are prevented from believing these things by experience (which shows us that those who profess to be Philosophers are often less wise and less reasonable than others who have never applied themselves to this study), I would have briefly explained, in that Preface, what all the knowledge which we now have consists in, and what stages of Wisdom we have reached. The first stage contains only notions so clear in themselves that they can be acquired without meditation. The second includes everything which the experiences of the senses make known to us. The third, what the conversation of other men teaches us. To these we can add, as the fourth, the reading of Books, not of all, but specifically of those written by persons capable of giving us good instruction; for reading is a sort of conversation which we have with the authors of those Books. And it seems to me that all the Wisdom which we ordinarily have is acquired in these four ways; for I am not including divine revelation here, since it does not lead us in stages but raises us up all at once to an infallible belief. Now there have always been great men who have attempted to find a fifth stage, incomparably higher and more certain than the other four, in order to attain Wisdom: this is the search for the first causes and the true Principles from which one might deduce the reasons for everything which we are capable of knowing; and it is especially to those men who have devoted their efforts to attaining this that the name 'Philosophers' has been given. Yet I do not know of any who have so far succeeded in this project. The first and principal Philosophers whose writings we have are Plato and Aristotle; between whom there was no difference except that the former, following in the footsteps of his teacher Socrates, ingenuously confessed that he had not yet been able to discover anything certain, and contented himself with writing the things which seemed to him likely; imagining for this purpose some Principles by which he attempted to give an explanation of the other things. Whereas Aristotle was less candid, and although he had been Plato's disciple for twenty years and had no Principles other than Plato's; he entirely changed the manner of expressing these Principles and propounded them as true and certain; even though it is unlikely that he ever judged them to be so. Now, these two men had much intelligence and much of the Wisdom which is acquired in the four ways previously listed; which gave them so much authority that those who came after them concentrated more on following their opinions than on seeking something better. And the principal dispute which their disciples had among themselves concerned whether one should question all things, or whether there were some which were certain. This led to foolish errors on both sides; for some of those who were in favor of doubt extended it even to the actions of life, so that they neglected to use prudence in their own conduct. And those who upheld certainty, supposing that it must depend upon the senses, entrusted themselves entirely to their senses; to such an extent that Epicurus is said to have dared to affirm, contrary to all the reasonings of Astronomers, that the Sun is no larger than it appears. A failing which can be observed in most disputes in which the truth is midway between the two opinions being expressed is that each man moves the further from the truth, the fonder he is of contradicting. However, the error of those who inclined too much toward doubt was not followed for long, and that of the others has been somewhat corrected in that it has been recognized that the senses deceive us in many things. Yet I do not know that the latter error has [yet] been entirely erased by a demonstration that certainty is not in the senses, but only in the understanding when it has evident perceptions; and that, while one has only the knowledge which is acquired through the first four stages of Wisdom, one must, as far as the conduct of life is concerned, neither doubt the things which seem true nor judge them to be so certain that one cannot change one's mind when obliged to do so by the evident nature of some reason. As a result of not having known that truth, or, if there were some who knew it, as a result of not having used it; most of those who have attempted to be Philosophers in recent centuries have blindly followed Aristotle in such a way as to often corrupt the sense of his writings by attributing to him diverse opinions which he would not recognize as his own if he returned to this world. And those who have not followed him (among whom were many of the best minds) were nevertheless immersed in his opinions in their youth (because these are the only ones taught in the schools), which prejudiced them to such an extent that they were unable to attain knowledge of the true Principles. And although I esteem all these men and do not wish to render myself odious by finding fault with them, I can support what I am saying with a proof which I do not believe any of them would disavow; which is that they all accepted as a Principle something of which they did not have perfect knowledge. For example, I know of no Philosopher who did not suppose that there was weight in terrestrial bodies. But although experience shows us clearly that the bodies which we call heavy descend toward the center of the earth, we do not on that account know the nature of what we call weight; that is to say, the cause or Principle which makes them thus descend, and we must learn it from elsewhere. The same can be said of the void and of atoms, and of heat and cold, of dryness, of humidity, of salt, of sulphur, of mercury, and of all similar things which some have taken as their Principles. Now all the conclusions which one deduces from a Principle which is not itself evident cannot be evident either, even though they may be deduced from it in an evident manner: from which it follows that all the reasonings which Philosophers based upon such Principles were unable to give them certain knowledge of anything, or, consequently, to take them one step forward in the pursuit of Wisdom. And if they found something true, it was only in some of the four ways enumerated above. Yet I do not wish to diminish in any way the honor which each of these men can claim. I am only obliged to say, for the consolation of those who have not studied, that, just as in travelling, so long as one turns one's back on the place to which one wishes to go, one moves further from it the longer and more rapidly one walks (so that even if one is subsequently put on the right road, one cannot arrive as quickly as if one had not previously walked); so, when one has bad Principles, the more one cultivates them and the more carefully one applies oneself to drawing various conclusions from them (thinking that to do so is to philosophize correctly), the further one gets from the knowledge of truth and from Wisdom. From this, it must be concluded that those who have learned the least about everything which has hitherto been called Philosophy are the most capable of learning true Philosophy.

After having made these things clearly understood, I should have wished to set down here the reasons which serve to prove that the true Principles by which one can reach that highest degree of Wisdom, in which consists the supreme good of human life, are those which I have propounded in this Book. And two reasons will be sufficient for that purpose, the first of which is that the Principles themselves are very clear, and the second is that all other things can be deduced from them: for only these two conditions are required of true Principles. Now, I easily prove that they are very clear; first, by the way in which I found them (that is, by rejecting all the things in which I could find the least reason for doubt); for it is certain that those things which could not thus be rejected after being examined attentively are the most evident and the most clear which the human mind can know. Thus, considering that he who wishes to doubt everything nevertheless cannot doubt that he exists while he is doubting, and that what reasons thus (being unable to doubt itself and yet doubting all the rest), is not what we call our body but what we call our soul or our mind; I took the being or the existence of that mind as the first Principle. From this I very clearly deduced the following: that there is a God who is the author of everything which is in the world; and who, being the source of all truth, did not make our understanding of a nature such that it could be mistaken in the judgment it makes of the things of which it has a very clear and very distinct perception. Those are all the Principles which I use concerning immaterial or Metaphysical things. And from those Principles, I very clearly deduce the Principles of corporeal or Physical things; namely that there are bodies extended in length, width, and depth, which have diverse figures and are moved in diverse ways. There, in short, are all the Principles from which I deduce the truth of other things. The other reason which proves the clarity of these Principles is that they have always been known, and indeed accepted as true and indubitable by all men; with the sole exception of the existence of God, which has been called into doubt by some because they attached too much importance to the perceptions of the senses and God can be neither seen nor touched. But although all the truths which I include among my Principles have always been known to everyone, there has nevertheless been no one up to this time, as far as I know, who has recognized them as the Principles of Philosophy, that is to say, as being such that one can deduce from them the knowledge of all the other things which are in the world. That is why it remains for me to prove here that they are such; and it seems to me that I cannot do so better than by demonstrating it from experience, that is, by urging my Readers to read this Book. For although I have not treated of all things in it, and although that would be impossible; I think I have explained all those which I have had occasion to consider in such a way that those who attentively read what I have written will have reason to be convinced that there is no need to seek Principles other than those which I have given, in order to attain all the highest knowledge of which the human mind is capable. This will especially be so if, after having read my writings, they take the trouble to consider how many diverse questions are explained there; and if, perusing the writings of others as well, they notice how few credible reasons others have been able to produce in an attempt to explain the same questions by means of Principles different from mine. And in order that they might undertake this more easily, I could have said to them that those who are thoroughly conversant with my opinions have much less difficulty in understanding and correctly evaluating the writings of others than do those who are not conversant with them; which is quite contrary to what I have just said about those who began with the former Philosophy, that is, that the more of it they have studied the less suited they usually are to correctly learning true Philosophy.

I would also have added a word of advice concerning the way to read this Book, which is that I would like it first to be read rapidly in its entirety, like a Novel, without the Reader forcing his attention too much or stopping at the difficulties which he may encounter in it, simply in order to have a broad view of the matters which I have treated in it. And after that, if the Reader judges that these matters merit examination, and is curious to know their causes; he can read the Book a second time, in order to notice the sequence of my reasonings. But again, he must not be discouraged if he cannot everywhere sufficiently follow the sequence or does not understand all my reasonings; it is only necessary to indicate with the stroke of a pen the places in which one finds some difficulty, and to continue to read to the end without interruption. If one then takes up the Book for a third time, I dare to believe that one will find in it the solution to most of the difficulties which one previously marked; and that if some still remain, one will finally find the solution to them by re-reading.

In examining the characteristics of many minds, I have noticed that there are almost none so unrefined or so slow that they would not be capable of attaining higher sentiments, and even of acquiring all the highest knowledge, if they were properly guided. And that can also be proved by reason: for, since the Principles are clear and we must deduce nothing from them except by very evident reasonings, we are always sufficiently intelligent to understand the things which depend upon them. But, in addition to the obstacle of prejudices (from which no one is entirely free, although they are most harmful to those who have studied bad science the most), it almost always happens that those who are of a cautious temperament neglect to study because they do not think themselves capable of it; while those who are more enthusiastic are over-hasty: as a result, people often accept Principles which are not evident, and draw from them conclusions which are not certain. That is why I should like to assure those who place too little trust in their abilities that there is nothing in my writings which they cannot completely understand, if they will take the trouble to examine them; and yet I should inform the others that even the most excellent minds will need much time and attention in order to notice all the things which it was my intention to include.

Next, in order to give a clear conception of the aim which I had in publishing my writings, I should wish to explain here the order which it seems to me ought to be observed so that one may learn. First, a man who thus far has only the common and imperfect knowledge which one can acquire in the four ways explained earlier must strive above all to form for himself a Moral code which can suffice to regulate the actions of his life; because that tolerates no delay, and because we must above all strive to live well. After that, he must also study Logic; not that of the schools, for that is properly speaking only a Dialectic which teaches the means of making others understand the things one knows, or even the means of speaking without judgment and at length about the things one does not know: as a consequence, that Logic corrupts rather than increases good sense. Rather, he must study that Logic which teaches how to use one's reason correctly in order to discover the truths of which one is ignorant; and because this depends greatly upon practice, he should drill himself for a long time by using the rules of Logic in relation to simple and easy questions, like those of Mathematics. Then, when he has become somewhat accustomed to discovering the truth in these questions, he must begin to apply himself seriously to true Philosophy, the first part of which is Metaphysics, which contains the Principles of knowledge; among which is the explanation of the principal attributes of God, of the immateriality of our souls, and of all the clear and simple notions which are in us. The second is Physics, in which, after having discovered the true Principles of material things, one examines, in general, the composition of the whole universe, and then, in particular, the nature of this Earth and of all the bodies which are most commonly found around it, like air, water, fire, the loadstone, and the other minerals. After this, it is also necessary to examine in particular the nature of plants, of animals, and above all, of man; in order to be capable of subsequently discovering all the other useful branches of knowledge. Thus, Philosophy as a whole is like a tree; of which the roots are Metaphysics, the trunk is Physics, and the branches emerging from this trunk are all the other branches of knowledge. These branches can be reduced to three principal ones, namely, Medicine, Mechanics, and Ethics (by which I mean the highest and most perfect Ethics, which presupposes a complete knowledge of the other branches of knowledge and is the final stage of Wisdom).

Now, just as it is not from the roots or from the trunk of trees that one gathers fruit, but only from the extremities of their branches, so the principal usefulness of Philosophy depends upon those parts of it which can only be learned last. But, although I am ignorant of almost all of those, the zeal which I have always had to strive to be of service to the public caused me to publish, ten or twelve years ago, some essays on the things which it seemed to me that I had learned. The first of these essays was a *Discourse on the Method of rightly conducting one's reason and seeking truth in the sciences*, in which I briefly stated the principal rules of Logic and of an imperfect Ethics, which one can follow provisionally while one still does not know anything better. The other essays were three treatises: the first *on Dioptrics*, the second *on Meteorology*, and the third *on Geometry*. In the *Dioptrics*, it was my intention to show that one can proceed far enough in Philosophy to achieve by its means a knowledge of those arts which are useful to life, because the designing of telescopes, which I explained there, is one of the most difficult tasks ever undertaken. In the *Meteorology*, I wished to make known the difference between the Philosophy which I study and that which is taught in the schools where it is customary to treat of the same subject. Finally, in the *Geometry*, I sought to demonstrate that I had discovered many things which were previously unknown and thus to provide grounds for believing that many others can still be discovered, in order to thereby incite all men to the search for truth. Subsequently, foreseeing the difficulty which many would have in conceiving the foundations of Metaphysics, I attempted to explain its principal points in a book of *Meditations* which is not very long, but whose length was increased and subject matter much illuminated by the objections concerning these *Meditations* which several very learned persons sent to me, and by the responses which I made to them. Then, finally, when it seemed to me that these preceding treatises had sufficiently prepared the minds of Readers to receive the *Principles of Philosophy*, I published these also; dividing the Book into four parts, the first of which contains the Principles of knowledge, which are what one can call first Philosophy or Metaphysics: that is why, in order to understand this first part well, it is appropriate to read beforehand the Meditations which I wrote on the same subject. The other three parts contain everything which is most general in Physics, namely, the explanation of the first laws or Principles of Nature, and the way in which the Heavens, the fixed Stars, the Planets, the Comets, and generally all the universe is composed; then, in particular, the nature of this earth, of air, of water, of fire, and of the loadstone, which are the bodies one can most commonly find everywhere about the earth; and of all the qualities which are observed in these bodies, such as light, heat, weight, and similar things: by which means I believe I have begun to explain all Philosophy in correct order, without having omitted any of those things which ought to precede the last things which I wrote. However, in order to pursue this project to completion, I ought hereafter to explain in the same way the nature of each of the other even more particular bodies which are on the earth, namely, minerals, plants, animals, and, principally, man. Finally, I ought to treat accurately of Medicine, Ethics, and Mechanics. That is what I would have to do in order to give men a perfectly complete body of Philosophy: and I do not yet feel so old; I do not have so little trust in my strength; I do not judge myself so far from the knowledge of what remains; that I would not dare to undertake to complete this project if it were possible for me to perform all the experiments which I would need in order to support and justify my reasonings. But, perceiving that to do this would require great expenditures which a private individual like myself could not meet without the help of the public, and not perceiving that I should expect that help; I believe I must henceforth content myself with studying for my personal instruction and that posterity will pardon me if I cease henceforth to work in its behalf.

However, in order that the extent to which I think I have already served posterity may be seen, I shall indicate here the benefits which I am convinced can be derived from my Principles. The first is the satisfaction which men will have in finding there many previously unknown truths; for although the truth often does not affect our imagination as much as falsehoods and shams, because it appears less admirable and more simple; yet the contentment which it gives is always more enduring and better founded. The second benefit is that by studying these Principles, men will gradually become accustomed to forming better judgments about all the things which they encounter, and thus to being Wiser. Thus, my Principles will have an effect contrary to that of the usual Philosophy; for it is easy to notice that that Philosophy makes those we call Pedants less capable of reason than they would be if they had never learned it. The third benefit is that the truths which my Principles contain, being very clear and very certain, will remove all subjects of dispute and will thus dispose minds to gentleness and harmony: quite contrary to the controversies of the schools, which imperceptibly render those who learn them more captious and more stubborn and are perhaps the first cause of the heresies and dissensions which now torment the world. The last and principal benefit of these Principles is that, by studying them, men will be able to discover many truths which I have not explained; and thus, by gradually passing from those already explained to new ones, will be able to acquire in time a perfect knowledge of all Philosophy and to ascend to the highest stage of Wisdom. For just as we see that although all the arts are crude and imperfect in the beginning, nonetheless, because they contain something true, whose effect is shown by experience; they gradually grow perfect with practice: so in Philosophy, when one has some true Principles, one cannot fail to encounter other truths from time to time by following them. And the falsity of Aristotle's Principles cannot be better proved than by saying that in the course of the many centuries for which men have followed them, no one has succeeded in making any progress by their means.

I very well know that there are minds which are so hasty and employ so little circumspection in what they do that they cannot build anything certain even when they have very solid foundations. And because it is usually they who are the promptest to write Books, they could quickly damage everything I have done, and introduce the kind of uncertainty and doubt into my way of philosophizing which I have carefully striven to banish from it, if their writings were taken to be mine or to be filled with my opinions. I have recently had an experience of this in a man who above all others was thought to wish to follow me, and of whom I had even written somewhere, "that I had so much trust in his mind, that I did not believe he had any opinion which I would not be willing to acknowledge as my own":[1] for last year he published a Book, entitled *Fundamenta Physicae*;[2] in which he seems to have said nothing about Physics and Medicine which he has not taken from my writings (both from those which I have published and from another still imperfect one concerning the nature of animals which came into his possession). However, because he transcribed badly, and changed the order, and denied certain truths of Metaphysics, upon which all Physics ought to be based; I am obliged to repudiate the Book entirely, and to beg Readers at this time never to attribute any opinion to me if they do not expressly find it in my writings, and not to accept any opinion in my writings or elsewhere as true, unless they very clearly see that it is deduced from true Principles.

I also well know that several centuries may pass before all the truths which can be deduced from these Principles have thus been deduced from them: because most of the truths which remain to be discovered depend upon certain specific observations, which will never be stumbled upon by chance but must be sought out with care and expense by very intelligent

---

[1] The person referred to here is Henricus Regius (Henri le Roy) who was a disciple of Descartes and professor of medicine at the University of Utrecht. Regius' views were bitterly opposed by Voetius, who was rector of the University. The quotation is a paraphrase of a passage in *A Letter from René Descartes to the most famous Man D. Gisbertus Voetius* (Amsterdam, 1643); see A. & T., VIII-2, 163.
[2] The actual title is *Fundamenta Physices* (Amsterdam, 1646).

men; and it will not easily happen that the same men who have the skill to put such experiments to good use will also have the possibility of performing them. Furthermore, most of the best minds have conceived such a poor opinion of Philosophy, because of the defects which they have noticed in that which has been practiced until now, that they will be unable to apply themselves to seeking a better one. But if at last both the difference which they will see between these Principles and all those of others, and the great sequence of truths which one can deduce from these Principles, indicate to them how important it is to continue the search for these truths, and to what degree of Wisdom, perfection of life, and joy these truths can lead; I dare to believe that there will be no one who will not strive to apply himself to such a profitable study, or who will not at least encourage and wish to assist with all his power those who fruitfully apply themselves to it. It is my wish that our descendents may see the success of this venture.

PART I

OF THE PRINCIPLES OF HUMAN KNOWLEDGE

PART I

1.   That whoever is searching after truth must, once in his life, doubt all things; insofar as this is possible.

Since we were born as children and made various judgments, {some good and some bad}, concerning perceptible[1] things before we had the complete use of our reason, we are diverted from a knowledge of the truth by many prejudices;[2] and it seems that we cannot be freed from these unless we attempt, once in our life, to doubt all those things in which we find even the slightest suspicion of uncertainty.

2.   That doubtful things must furthermore be held to be false.

Indeed, it will also be useful to consider as false those things {in} which we {can imagine the least} doubt; so that we may the more clearly discover those things which are the most certain and most easy to know.

3.   That this doubt is not meanwhile to be adopted for the conduct of life.

But for the time being, this doubt is to be limited solely to the contemplation of the truth. For where the conduct of life is concerned, we are not infrequently forced to accept what is only probable, or sometimes to choose one of two alternatives even though one may not appear more probable than the other; because very often the opportunity to act would pass before we could free ourselves from {all} our doubts.

4.   Why we can doubt perceptible things.

And since we are now only concerned with seeking the truth, we shall

---

[1] Latin: '*res sensibiles*'; 'sensible things'. Descartes normally uses '*sentio*', 'sense' or 'observe' for sense-perception and '*percipio*' or '*deprehendo*' for intellectual perception. Henceforth, 'perceptible' will refer only to sense-perception, and '*percipio*' will always be translated as 'perceive'.

[2] Latin: '*praejudicium*'; 'a hasty judgment'. Such judgments may be true; see Part II, Article 1.

begin by doubting whether any perceptible or imaginable things exist: first
because we perceive that our senses sometimes err, and it is prudent never
to place too much trust in whatever has even once deceived us; and next
because every night in our dreams we seem to observe or to imagine
innumerable things which are non-existent {elsewhere}; and to a man thus
doubting, there appear no indications by which he can distinguish sleep
from waking with certainty {and can know whether the thoughts which
come in dreams are more false than the others}.

5.   Why we can even doubt Mathematical demonstrations.

We shall also doubt the remaining things which we formerly held to be
most certain; even Mathematical demonstrations and even those principles
which until now we thought to be known of themselves:[3] both because we
see that sometimes some men have erred in such things, and have accepted
as very certain and self-evident things which seemed to us false; and above
all, because we have heard that there is a God who can do all things, and by
whom we were created. For we do not know whether He chose to make us in
such a way that we are always mistaken, even about those things which
appear to us to be the best known of all; because it seems as possible that
this could have occurred as that we should sometimes err, which, as we have
already noted, does occur. And if we imagine ourselves to exist, not as a
result of [an act of] a most powerful God, but either of ourselves, or of
any other thing: the less powerful we consider the author of our origin, the
more credible it will be that we are so imperfect that we are always
mistaken.

6.   That we have the free will to withhold our assent in doubtful
             matters, and thus to avoid error.

However, by whomever we may have been created, and however
powerful and however deceitful he may be; we nonetheless experience in
ourselves a freedom such that we can always abstain from believing those
things which are not absolutely certain and established; and thereby always
avoid error.

7.   That it is not possible for us to doubt that, while we are
             doubting, we exist; and that this is the first thing which we know
             by philosophizing in the correct order.

[3] Latin: 'per se nota'; 'known by means of themselves' or 'self-evident'.

Further, while rejecting in this way all those things which we can somehow doubt, and even imagining them to be false, we can indeed easily suppose that there is no God, no heaven, no material bodies; and even that we ourselves have no hands, or feet, in short, no body; yet we do not on that account suppose that we, who are thinking such things, are nothing: for it is contradictory for us to believe that that which thinks, at the very time when it is thinking, does not exist. And, accordingly, this knowledge,[4] *I think, therefore I am*,[5] is the first and most certain to be acquired by and present itself to anyone who is philosophizing in correct order.

8.   That from this we understand the distinction between the soul[6] and the body, or between a thinking thing and a corporeal one.

And {it seems to me that} this is the best path to the understanding of the nature of the mind, and of the distinction between the mind[7] and the body. For in examining what we may be, while supposing all things different from ourselves {and outside our mind} to be false; we clearly perceive that extension, or figure, or local motion (or any similar thing which must be attributed to a body) does not belong to our nature, but only the faculty of thinking, which is therefore known prior to and more certainly than any corporeal things; for we have already perceived this [thinking], and yet are still doubting the rest.[8]

9.   What thought is.

By the word 'thought', I understand all those things which occur in us while we are conscious, insofar as the consciousness of them is in us. And so not only understanding, willing, and imagining, but also sensing, are here the same as thinking. For if I say, I see, or I walk, therefore I am; and if I deduce[9] this [conclusion] from seeing or from walking which is performed by the body; the conclusion is not absolutely certain: because (as often happens in dreams) I can think that I am seeing or walking, even though I

---

[4] The French has "conclusion" here.

[5] Latin: '*ego cogito, ergo sum*'.

[6] Latin: '*anima*'; 'soul' (as distinct from body), cf. Part IV, note 123.

[7] Latin: '*mens*'; 'mind'.

[8] The French has "...; since we are still doubting whether any body exists in the world, and yet we know with certainty that we think." here.

[9] Latin: '*intellego*'; 'understand', 'deduce', or 'recognize'.

may not open my eyes, and may not be moved from my place; and indeed, even though I may perhaps have no body. But if I deduce this from {the action of my mind, or} the very sensation or consciousness of seeing or of walking; the conclusion is completely certain, for it [the premise] then refers to the mind which alone perceives or thinks that it is seeing or walking.

10.   That those things which are simplest and known of themselves are rendered more obscure by Logical definitions; and that such things are not to be included among knowledge acquired through study.[10]

I am not explaining here many other terms which I have already used, or which I shall use in what follows, because they seem to me sufficiently known of themselves. And I have often noticed that Philosophers have erred in striving to explain by Logical definitions those things which were simplest and known of themselves; for in this way, Philosophers rendered them more obscure. And when I stated that this proposition, *I think, therefore I am*, was the first and most certain of all which would present themselves to anyone who was philosophizing in correct order, I did not on that account deny that it was previously necessary to know what thought, existence, and certainty are; and similarly to know that it is impossible that that which thinks does not exist, and such; but because these are very simple notions which by themselves do not provide knowledge of any existing thing,[11] I consequently did not judge that they should be enumerated here.

11.   How our mind is better known than the body.

But now, in order to understand that our mind is not only known earlier and more certainly than the body, but also more clearly; it must be noted that it is very well known by the natural enlightenment[12] {which is in our souls} that no properties or qualities belong to nothingness; and that accordingly, wherever we perceive some properties or qualities, there we must necessarily find a thing or substance to which they belong; and that

[10] The French title is: "That there are notions so clear in themselves that one renders them obscure by attempting to define them in the scholastic manner, and that these notions are not acquired through study, but are born with us."

[11] "... knowledge of the existence of any thing," may be intended here.

[12] Latin: '*lumen*'; 'light', 'enlightenment', or 'understanding'.

the more properties or qualities we perceive in the same thing or substance, the more clearly we know it. However, it is obvious that we perceive more properties or qualities in our mind than in any other thing; since absolutely nothing can cause us to know something other than our mind, without at the same time bringing us with much more certainty to a knowledge of our mind itself. For example, if I judge that the earth exists from the fact that I touch or see it; from that very fact, I must be still more persuaded that my mind exists: for it can perhaps happen that I judge that I touch the earth, although no earth exists; but it cannot be the case that I make this judgment and that my mind which makes this judgment is nothing; and the same for the remaining [judgments]. {And all the other things which enter our minds enable us to reach the same conclusion: that we who are thinking them, exist; even though they may be false or have no existence}.

12.   Why this does not become equally well known to everyone.

And this seemed otherwise to those who did not philosophize in the correct order, solely because they never distinguished the mind from the body with sufficient care. And although they thought that they were more certain that they themselves existed than that anything else did, they did not however notice that by 'themselves', only their minds should have been understood in that context {where Metaphysical certainty was concerned}. On the contrary, they instead understood by 'themselves' only their bodies, which they saw with their eyes and touched with their hands, and to which they incorrectly attributed the power of sense-perception; and this prevented them from perceiving the nature of the mind.

13.   In what sense {it is possible to say that} the knowledge of
              remaining things depends on a knowledge of God.

Moreover, when the mind, which {thus} knows itself but is still doubting all other things, looks around on every side in order to extend its knowledge further: it first discovers in itself the ideas of many things; and as long as it is only contemplating these ideas and neither affirming nor denying that there is anything similar to them outside itself, it cannot err. The mind also discovers [in itself] certain common notions,[13] and forms various proofs from these; and as long as it is concentrating on these proofs it is entirely

---

[13] Latin: 'communes notiones'; 'common notions' or 'universal conceptions.

convinced that they are true. Thus, for example, the mind has in itself the ideas of numbers and figures, and also has among its common notions, *that if equals are added to equals, the results will be equal,* and other similar ones; from which it is easily proved that the three angles of a triangle are equal to two right angles, etc. Accordingly, the mind is convinced that these and similar things are true as long as it is considering the premises from which {and the order in which} it has deduced these things. But because it cannot always consider these premises, when it later {remembers some conclusion without attending to the order in which it can be demonstrated and} remembers that it does not yet know whether it was perhaps created of such a nature that it errs even in those things which appear most evident to it; the mind sees that it rightly doubts such things, and cannot have any certain knowledge until it has come to know the author of its origin.

14.   That from the fact that necessary existence is contained in our conception of God, it is properly concluded that God exists.

Next, considering that among the diverse ideas which the mind has in itself there is one of a being who is supremely powerful, omniscient, and perfect in the highest degree, and that this is by far the most exceptional[14] of all; the mind understands that entirely necessary and eternal existence is contained in this idea rather than the merely possible and contingent existence which is contained in the ideas of all other things which it distinctly perceives. And just as, for example, the mind is entirely convinced that a triangle has three angles which are equal to two right angles, because it perceives that the fact that its three angles equal two right angles is necessarily contained in the idea of a triangle: so, solely because it perceives that necessary and eternal existence is contained in the idea of a supremely perfect being, the mind must clearly conclude that a supremely perfect being exists.

15.   That necessary existence is not similarly contained in the concepts of other things, but only contingent existence.

And the mind will be more convinced of this, if it attends to the fact that it finds in itself the idea of no other thing in which it notices necessary existence to be similarly contained. For from this it will understand that

[14] Latin: '*praecipuus*'; 'most important', 'most exceptional', or 'most fundamental'. The French text omits the entire clause.

this idea of a supremely perfect being has not been devised by it, and does not present some imaginary nature, but {that it is imprinted on the mind by} a true and immutable nature, and one which cannot fail to exist; since necessary existence is contained within it.

16.   That prejudices prevent this necessity of God's existence from being clearly known by all.

I say that our mind will easily believe this,[15] if it has first entirely freed itself from prejudices. But because we have been accustomed to distinguishing essence from existence in all other things, and also to arbitrarily imagining various ideas of things which do not, or did not, exist; it easily happens that when we are not completely intent upon the contemplation of the supremely perfect being, we may wonder whether that idea is perhaps one of those which we have arbitrarily imagined, or at least one of those to whose essence existence does not belong.

17.   That the greater the objective perfection of each of our ideas, the greater [the perfection of] its cause must be.

Moreover, upon considering further the {various} ideas which we find in ourselves, we see that insofar as they are particular modes of {our} thinking, they do not differ much from one another, but that insofar as one represents one thing, and another another thing, they are very different. And [we see] that the more objective perfection[16] they contain, the more perfect their cause must be. For instance, if someone has in himself the idea of some very ingenious machine, it can properly be asked from what cause

[15] Descartes uses expressions like 'it is easy to believe....,' 'we will easily understand...,' etc. throughout the *Principles*. In a letter written to Mersenne in 1637, concerning a doubt which Fermat had expressed about the *Dioptrics*, Descartes says: "But I believe he formed this doubt because he imagined that I myself was doubtful on this matter; and because (since I stated on p. 8, line 24: *For it is very easy to believe that the inclination to motion must in this follow the same laws as movement [itself]*) he thought that when I said that a thing is easy to believe, I meant that it is merely probable. In this, he has greatly mistaken my view. For I consider everything which is merely likely almost as false; and when I say that a thing is easy to believe, I do not mean that it is only probable, but that it is so clear and evident that it is not necessary for me to stop to demonstrate it.": A. & T., I, 450–451.

[16] The objective perfection of an idea is the amount or degree of perfection which the *object* of that idea is conceived of or represented as possessing. The formal perfection of a thing is the actual degree of perfection it possesses. A cause which possesses perfection of a different kind from its effect is said to possess the perfection of that effect "eminently".

he obtained that idea: that is, whether he saw somewhere such a machine made by someone else; or whether he learned the mechanical sciences so perfectly or there is so much power in his mind that, without ever having seen it anywhere, he was able to devise it himself. For all the ingenuity which is contained in that idea only objectively, or as if in a picture, must be contained in the cause of that idea, whatever that cause may be, not merely objectively or representatively, but in fact formally or {even more} eminently, at least in the first and principal cause [of that idea].

18.   That from this it is once again concluded that God exists.

Thus, because we have in us the idea of God or of a supreme being, we can justly examine from what cause we have this idea; and we shall find so much immensity {of perfection} in this idea, that we shall thereby be completely certain that it cannot have been imparted to us except by a thing in which there truly was a complete complement of all perfections, that is, except by a God who really exists. For it is very well known from [our] natural enlightenment, not only that nothing cannot produce anything; but also that that which is more perfect is not produced by an efficient and total cause which is less perfect; and moreover that there cannot be in us the idea or image of any thing, of which there does not exist somewhere (either in us or outside us), some Original, which truly contains all its perfections. And because we in no way find in ourselves those supreme perfections of which we have the idea; from that fact alone we rightly conclude that they exist, or certainly once existed, in something different from us; that is, in God: and from this {and the fact that they were infinite} it most evidently follows that they still exist.

19.   That even though we do not comprehend the nature of God, nevertheless, His perfections are known to us more clearly than any other thing.

And this is sufficiently certain and evident to those who have been accustomed to contemplating the idea of God and to noticing His supreme perfections. For although we do not comprehend these perfections, because of course it is of the nature of the infinite not to be comprehended by us who are finite, we can however understand them more clearly and more distinctly than any corporeal things; because they fulfil our mind more, and are more simple, and are not obscured by any limitations.

{Further, there is no contemplation which can be more helpful in perfecting our understanding or is more important than this one, since to consider an object which has no limits to its perfections fills us with satisfaction and assurance}.

20.   That we were not created by ourselves, but by God, and that consequently He exists.

However, because not all men notice this, and also because, while those who have the idea of some ingenious machine usually know whence they obtained that idea; we do not similarly remember that at some particular time the idea of God came to us from Him, because we have always had it: it still remains to be asked by whom we ourselves, who have the idea of the supreme perfections of God, were created. For certainly it is very well known from natural enlightenment that whatever knows something more perfect than itself, does not [come to] exist by means of itself: for it would {by the same means} have given itself all the perfections of which it has the idea; and that accordingly it cannot exist as a result of anything which does not have in it all those perfections, that is to say, which is not God.

21.   That the duration of our existence suffices to prove the existence of God.

And nothing can obscure the clarity of this proof, at least if we consider the nature of time or of the duration of things; which is such that its parts do not depend upon one another, or ever exist simultaneously; and that, accordingly, from the fact that we now exist, it does not follow that we shall also exist a moment from now, unless some cause (that is, the same one as that which first produced us) continually produces us, as it were, anew; that is, conserves us.[17] For we easily understand that there is in us no power by which we may conserve ourselves; and that He in whom there is so much power that He can conserve us separately from Himself, must also conserve Himself all the more, or rather, must require no conservation by anyone, and finally, must be God.

---

[17] The underlying assumption here is that unless one thing implies another, it cannot be the cause of the other; that is, that all causal connections are, ultimately, necessary connections.

22.   That from our way of knowing the existence of God, all of His
              attributes which are discernible by the power of natural
              understanding are at the same time known.

Furthermore, there is a great advantage in this way of proving the
existence of God, i.e., by the idea of Him: we simultaneously learn what He
is, to the extent permitted by the weakness of our nature. For of course,
while examining the idea of Him which is innate in us, we see that He is
eternal, omniscient, omnipotent, the source of all goodness and truth, the
creator of all things, and, finally, that He has in Him all those things in
which we can clearly observe some perfection which is infinite or limited by
no imperfection.

23.   That God is not corporeal, and does not perceive by the senses
              as we do, and does not will the fault of error.[18]

For there are very many things in which, although we may recognize
some perfection, we however also perceive some imperfection or
limitation; and which accordingly cannot belong to God. Thus, because
divisibility is incorporated together with local extension in the nature of
corporeal things, and because it is an imperfection to be divisible, it is
certain that God is not corporeal. And although, in us, it is a certain
perfection for us to perceive by sense, however, because all sensation
involves our being acted upon, and because this sufferance is dependence
upon something; it must be thought that God in no way perceives by sense
but simply understands and wills: and that He does not do this as we do, by
operations which are in some way different from one another; but in such
a way that by a single, always identical, and very simple action, He
simultaneously understands, wills, and performs everything. I say
"everything", that is, all things: for He does not will the fault of error, since
it is not a thing.

24.   That the knowledge of created things is reached from the
              knowledge of God, by recollecting that He is infinite and that
              we are finite.

But now, because God alone is the true cause of all things which are or
can be, it is obvious that we shall be following the best method of

---

[18] Latin: *'peccatum'*; 'sin', 'error', or 'mistake'.

philosophizing if we strive to deduce the explanation of the things created by Him from the knowledge of God Himself {and the notions innate in us}; so that we may thus acquire the most perfect knowledge, which is that of effects through their causes. In order to undertake this with sufficient caution and without risk of erring, we must take the precaution of always remembering as clearly as possible both that God the creator of things is infinite, and that we are in every way finite.

25.   That all things which have been revealed by God must be believed, although they may surpass our power of comprehension.

Thus, if it happens that God reveals to us something, concerning Himself or other things, which exceeds the natural powers of our understanding[19] (as the mysteries of the Incarnation and the Trinity already do); we shall not refuse to believe those things, although we {perhaps} do not clearly understand them. And we shall not wonder in the slightest that there are many things, both in His boundless nature and also in the things created by Him, which surpass our power of comprehension.

26.   That we must never discuss the infinite, but must simply consider those things in which we notice no limits as indefinite; as, for instance, the extension of the world, the divisibility of parts of matter, the number of stars, etc.

Thus we shall never be wearied by any debates concerning the infinite. For of course, inasmuch as we are finite, it would be absurd for us to attempt to determine anything concerning the infinite, and thus {suppose it finite by an} attempt as it were to prescribe limits to it and comprehend it. Therefore, we shall not bother to respond to those who ask whether half of a given infinite line would also be infinite; or whether infinite number is even or odd, and such: because surely only those who judge their own mind to be infinite ought to think about such things. Moreover, we shall not affirm that all those things in which we have been able to find no limit after some consideration are infinite; but shall view them as indefinite. Thus, because we cannot imagine an extension so great that we do not understand

---

[19] The French text reads: "So, if He grants us the blessing of revealing, either to us or to some others, things which exceed the ordinary reach of our mind, ..."

that a still larger one can exist; we shall say that the magnitude of possible things is indefinite. And because a body cannot be divided into so many parts that these individual parts are not understood to still be divisible; we shall think that quantity is indefinitely divisible. And because it is not possible to imagine such a great number of stars that we do not believe that God could have created still more, we shall suppose their number to also be indefinite; and similarly for the rest.

27.   What the difference between 'indefinite' and 'infinite' is.

And we shall say that these things are indefinite rather than infinite: both in order to reserve the term 'infinite' for God alone; because in Him alone, in every respect, we not only recognize no limits {to His perfection}, but also in a positive sense understand that there are none; and also because we do not similarly understand in a positive sense that other things are in some respect without limits, but only in a negative sense acknowledge that we cannot find their limits if they have any. {And thus we know that these things are not absolutely perfect, because we understand that this apparent lack of limits results from the weakness of our understanding rather than from the nature of these things}.

28.   That we must not examine the final causes of created things, but rather their efficient causes.

And so, finally, concerning natural things, we shall not undertake any reasonings from the end which God or nature set Himself in creating these things, {and we shall entirely reject from our Philosophy the search for final causes}: because we ought not to presume so much of ourselves as to think that we are the confidants of His intentions. But, considering Him as the efficient cause of all things, we shall see what the natural enlightenment with which He endowed us reveals must be concluded (concerning those of His effects which appear to our senses), from those of His attributes of which He willed that we should have some notion.[20] We shall however be

---

[20] The French text reads: "But considering Him as the Author of all things, we shall only attempt to discover, by means of the faculty of reasoning which He has placed in us, how those things which we perceive by the intermediary of our senses can have been created; and we shall be assured, by those of His attributes of which He wished us to have some knowledge, that that which we have once clearly and distinctly perceived to belong to the nature of those things has the perfection of being true." Also, the French omits the final sentence of the article.

mindful that, as has already been said, we must only trust in this natural enlightenment for as long as nothing contrary is revealed by God Himself.

29.   That God is not the cause of errors.

The first attribute of God which comes into consideration here is that He is veracious in the highest degree, and the giver of all understanding: so that it would be completely contradictory for Him to deceive us, or to be specifically and positively the cause of the errors to which we know from experience that we are subject. For although the ability to deceive may perhaps seem among us men to be some proof of cleverness, the will to deceive certainly never proceeds from anything other than malice, or fear, or weakness; and, consequently, cannot occur in God.

30.   That it follows from this that all the things which we clearly perceive are true, and that the doubts previously listed are removed.

And from this it follows that the natural enlightenment or the faculty of knowing given to us by God, can never attain any object which is not true, insofar as it is clearly and distinctly perceived. For He would deservedly have to be called a deceiver if He had given us a faculty which was perverse and which mistook the false for the true {when we were using it correctly}. Thus is removed the greatest doubt, which proceeded from the fact that we did not know whether we were perhaps of such a nature that we were deceived even in those things which seemed to us to be the most evident. Indeed, all the other causes of doubt, previously listed, are easily removed by this principle. Thus, Mathematical truths must no longer be mistrusted by us, since they are most manifest. And if we notice something which is clear and distinct in our sensations (either while we are awake or while we are asleep), and we distinguish it from what is confused and obscure; we shall easily recognize, in anything whatever, what should be taken to be true. Nor do we need to pursue these matters here at greater length, since they have already been treated in one way or another in the *Metaphysical Meditations*, and since a more precise explanation of them depends on a knowledge of what follows.

31.   That our errors are only negations if considered in relation to God; but that, considered in relation to us, they are privations.

But even though God is not a deceiver; because it nevertheless often happens that we are deceived, in order to investigate the origin and cause of our errors and to learn to prevent them, it must be noticed that they do not depend so much upon the understanding as upon the will; and that they are not things whose creation requires the real participation of God. Rather, when our errors are considered in relation to Him, they are only negations; {that is, He did not give us everything which He could have and which . . . He was not obliged to give us}. And when they are considered in relation to us, they are privations {and imperfections}.

32.   That there are only two modes of thinking in us; that is, the
           perception of the intellect and the operation of the will.

Of course, all the modes of thinking which we experience in ourselves can be reduced to two general ones: the first of which is perception, or the operation of the intellect; while the second is volition, or the operation of the will. For sense-perception, imagining, and pure understanding, are only diverse modes of perceiving; and desiring, having an aversion, affirming, denying, and doubting, are diverse modes of willing.

33.   That we do not err except when we judge of a thing which is
           insufficiently perceived.

Moreover, when we perceive something, it is obvious that we do not err provided only that we affirm or deny absolutely nothing about that thing; any more than we err when we affirm or deny only those things which we clearly and distinctly perceive ought to be thus affirmed or denied: rather, we only err when (as occurs) we make a judgment about something even though we do not perceive it correctly.

34.   That not only the understanding but also the will is required in
           order to judge.

And in order to judge, the understanding is required (because we can make no judgment about a thing which we in no way perceive); but the will is also required, in order that assent may be given to the thing which has been perceived in some way. Moreover, complete perception of the thing is not required, at least not in order to judge [it] in some way or another; for we can assent to many things which we know only very obscurely and confusedly.

35.   That the will extends further than the understanding, and thus
           is the cause of error.

And indeed the perception of the understanding is extended only to those
few things which are presented to it, and is always very finite. But the will
can, in a certain way, be said to be infinite; because we never notice
anything which can be the object of some other will (or of that boundless
will which is in God), to which our will does not also extend itself: thus we
can easily extend it beyond those things which we clearly perceive; and
when we do this, it is not surprising that we happen to err.

36.   That our errors cannot be imputed to God.

However, we cannot in any way imagine, because of the fact that God
did not give us an omniscient understanding, that He is the author of our
errors. For it is of the nature of created understanding that it should be
finite; and of the nature of finite understanding that it should not extend
itself to all things.

37.   That the highest perfection of man is that he acts freely, or
           through the will; and that this makes him worthy of praise or
           blame.

But that the will should extend very widely is also in accordance with its
nature; and it is the highest perfection in man that he acts through the will,
that is, freely, and thus in a certain way he alone is the author of his actions,
and deserves praise on account of them {when he conducts them well}. For
machines are not praised because they perform perfectly all the movements
for which they were constructed, because they necessarily perform those
movements in that way; however, their maker is praised for having made
them so perfect, because he did not make them necessarily, but freely. In the
same way, we must certainly be given more credit for voluntarily embracing
the truth, when we do embrace it, than if we were unable not to embrace it
{and were forced into it by a principle foreign to us}.[21]

---

[21] On this point, cf. Article 43, where Descartes seems to claim that the will is powerless to
withhold assent from clear and distinct perceptions.

38.   That when we err, it is a weakness in our action, not in our
             nature; and that the guilt of subordinates can often be
             attributed to other masters, but never to God.

However, our falling into error is a weakness in our action or in the use of
our freedom, but not in our nature; because our nature is the same when we
judge incorrectly as when we judge correctly. And although God could
have placed so much discernment in our understanding that we would
never err, we however have no right to demand this of Him. And although,
among men, if someone has the power to prevent some evil and yet does not
prevent it, we say that he is the cause of that evil: however, we must not in
the same way think that God is the cause of our errors on the grounds that
He could have brought it about that we would never err. For the power
which some men have over others is instituted so that they may use it to
restrain others {inferior to themselves} from evil; whereas the power which
God has over all men is as absolute and free as possible: accordingly, we
ought to give Him deepest thanks for the blessings which He has bestowed
upon us; but we have no right to complain because He did not bestow all
the things which we know He could have {perhaps} bestowed.

39.   That freedom of the will is known of itself.

Further, it is so manifest that there is freedom in our will, and that we
have complete power to either assent or not assent to many things, that this
must be numbered among the first and most common notions innate in us.
And this was very obvious a little earlier, when we were striving to doubt
everything and had gone so far as to imagine that some very powerful
author of our origin was attempting to deceive us in every way; for we
nevertheless experienced in ourselves the freedom to be able to abstain
from believing those things which were not absolutely certain and
confirmed. For no things can ever be better known of themselves and
better proved than those which did not seem doubtful at that time.

40.   That it is also certain that all things are pre-ordained by God.

But because, now knowing God; we perceive in Him a power so limitless
that we would think it a sin to judge that we could ever do anything which
He had not previously pre-ordained: we can easily involve ourselves in
great difficulties if we attempt to reconcile God's pre-ordaining [everything]
with the freedom of our will, and to understand both at the same time.

41.   How the freedom of our will and God's pre-ordaining may be reconciled.

However, we shall rid ourselves of those difficulties, if we remember that our mind is finite; whereas the power of God (by means of which He has not only had fore-knowledge from eternity of all things which are or can be but has also willed and pre-ordained them) is infinite: and that therefore we apprehend this power sufficiently to perceive clearly and distinctly that it is in God; but do not understand it sufficiently to see by what means it leaves the free actions of men undetermined; and if we remember that, even so, we are so conscious of the freedom and indifference which are in us, that there is nothing which we understand more evidently and perfectly. For, simply because we do not understand a thing which we know must of its nature be incomprehensible to us; it would be absurd to doubt other things which we understand inwardly and experience in ourselves.

42.   How, although we are unwilling to err, we nonetheless err through our will.

But now, since we know that all our errors depend upon the will, it may seem surprising that we ever err; because there is no one who wills to err. But to will to err is very different from willing to assent to those things in which it happens that an error is found. And although truly there is no man who expressly wills to err, there is hardly any man who does not often will to assent to those things in which, unknown to him, an error is contained. Indeed, the very desire to pursue the truth frequently causes those who do not know the correct method of pursuing it to make a judgment about those things which they do not {sufficiently} perceive, and on that account to err.

43.   That we never err when we assent only to things which are clearly and distinctly perceived.

However, it is certain that if we give assent only to those things which we clearly and distinctly perceive, we will never accept anything false as being true. I say that it is certain because, since God is not a deceiver, the faculty of perceiving which He gave us cannot lead toward what is false; any more than can the faculty of assenting when it is extended only to those things which are clearly perceived. And even if this were not proved by any

reasoning, it is impressed by nature upon the minds of all; so that whenever
we clearly perceive something, we spontaneously assent to it and cannot in
any way doubt that it is true.

44.   That we always judge badly when we assent to things which are
              not clearly perceived, even if by chance we stumble upon the
              truth; and that this sometimes happens because we are
              supposing these things to have formerly been adequately
              perceived by us.

It is also certain that when we assent to some judgment which we do not
[clearly] perceive, we either err, or we stumble upon the truth only by
chance and thus do not know that we are not in error. But it very rarely
occurs that we assent to those things which we notice have not been
perceived by us: because natural enlightenment dictates to us never to
judge of a thing unless it is known. However, we very frequently err in that
there are many things which we think that we formerly perceived; and after
they have been committed to memory, we assent to these things as if they
had been thoroughly perceived; when in fact we never truly perceived them.

45.   What a clear perception is, and what a distinct one is.

Indeed, in their whole lives, many men never perceive anything whatever
accurately enough to make a sure judgment about it; because a perception
upon which a sure and unquestionable judgment can rest must not only be
clear, it must also be distinct. I call 'clear' that perception which is present
and manifest to an attentive mind: just as we say that we clearly see those
things which are present to our intent eye and act upon it sufficiently
strongly and manifestly. On the other hand, I call 'distinct', that perception
which, while clear, is so separated and delineated from all others that it
contains absolutely nothing except what is clear.

46.   That, by the example of pain, it is shown that a perception can
              be clear even though it is not distinct; but that it cannot be
              distinct unless it is clear.

Thus, when someone feels some great pain, the perception of pain is
indeed very clear in him, but is not always distinct; for commonly men
confuse that perception with their uncertain {false} judgment about its

nature; because they believe something resembling the feeling of pain to be in the painful part. And thus a perception which is not distinct can be clear; but no perception can be distinct unless it is clear.

47.   That in order to correct the prejudices of our youth, we must consider [our] simple notions, and what is clear in each one.

And indeed, in our early youth, our mind was immersed in the body in such a way that although it perceived many things clearly, it however never perceived anything distinctly; and since it nevertheless made judgments about many things at that time, we acquired as a result many prejudices which most men never subsequently discard. However, in order that we may free ourselves from these, I shall here briefly enumerate all the simple notions of which our thoughts are composed; and in each one, I shall distinguish what is clear, from what is unclear or misleading.

48.   That all objects which come within our perception are to be regarded as things, or as states of things, or as eternal truths; and an enumeration of the things.

And we consider whatever objects come within our perception either as things, or as certain states of things; or else as eternal truths which have no existence outside our thought. Of those which we consider as things, the most general are *substance, duration, order, number*, and others of this sort (if any), which extend to all kinds of things. However, I do not recognize more than two principal kinds of things: one is intellectual or cogitative things, that is, things pertaining to the mind or to thinking substance; and the other, material things, or things pertaining to extended substance or body. Perception, volition, and all modes of perceiving and willing pertain to thinking substance; while size (or extension in length, width, and depth), figure, motion, situation, divisibility of its parts, and such, pertain to extended substance. However, we also experience in ourselves certain other things which should be attributed neither solely to the mind nor solely to the body, and which, as I shall show later in the proper place,[22] originate from the close and profound union of our mind with the body: specifically, the appetites of hunger, thirst, etc.; and similarly the emotions or passions of the soul (which do not consist solely in thought), for example, the

[22] See Part IV, Articles 189–191.

emotions of anger, merriment, sadness, love, etc.; and finally all sensations, such as pain, pleasure, light and color, sounds, odors, tastes, heat, hardness, and the other tactile qualities.

49.   That eternal truths cannot be thus enumerated, but that there is no need.

And we consider all these as things, or qualities or modes of things. However, when we acknowledge that it is impossible for something to be made out of nothing, then this proposition: *Nothing is made from nothing*, is not considered to be some existing thing, or even to be the mode of a thing, but a certain eternal truth which resides in our mind, and is called a common notion, or an axiom. The following propositions also are of this type: *It is impossible for the same thing to be and not to be at the same time: What has been done cannot be undone: He who thinks cannot not exist while he is thinking*: as are innumerable others which of course cannot easily all be listed, but neither can they fail to be known when the occasion to think of them occurs and when we are not blinded by any prejudices.

50.   That these eternal truths are clearly perceived, but not all [are perceived] by all men, because of prejudices.

And indeed, as far as these common notions are concerned, there is no doubt that they can be clearly and distinctly perceived, for otherwise they would not deserve to be called common notions. Nor is there any doubt that certain of these notions are in fact not equally worthy of this name where all men are concerned, because they are not equally perceived by all. However, I do not think that this is because the faculty of knowing extends further in some men than it does in others; but because these common notions may be opposed by the prejudiced opinions of some men, who consequently cannot easily grasp these notions: although those others who have freed themselves from these prejudices, perceive them most evidently.

51.   What substance is, and that this term does not apply univocally to God and to created things.

However, as for what we regard as things or as modes of things, it is worthwhile for us to consider them individually and separately here {in order to distinguish what is obscure from what is evident in our notion of them}. By '*substance*', we can understand nothing other than a thing which exists in such a way that it needs no other thing in order to exist. And indeed only one substance which needs absolutely no other thing can be understood; i.e., God. We perceive that, on the contrary, all others can exist only with the aid of God's participation. And consequently the term 'substance' does not apply to God and to those other things "univocally" (as is customarily {and rightly} said in the Schools), that is, no meaning of this term can distinctly be understood which is common to God and to created things. {But because, among created things, some are such that they cannot exist without some others; we distinguish them from those which require only the normal participation of God by naming the latter substances and the former the qualities or attributes of these substances}.

52.   That the term 'substance' is univocally applicable to mind and body; and how substance is known.

However, corporeal substance and created mind, or thinking substance, can be understood from this common concept: that they are things which need only the participation of God in order to exist. Yet substance cannot be initially perceived solely by means of the fact that it is an existing thing, for this fact alone does not *per se* affect us;[23] but we easily recognize substance from any attribute of it, by means of the common notion that nothingness has no attributes and no properties or qualities. For, from the fact that we perceive some attribute to be present, we {rightly} conclude that some existing thing, or substance, to which that attribute can belong, is also necessarily present.

53.   That each substance has one principal attribute, thought, for example, being that of mind, and extension that of body.

And substance is indeed known by any attribute [of it]; but each substance has only one principal property which constitutes its nature and essence, and to which all the other properties are related. Thus, extension in length, breadth, and depth constitutes the nature of corporeal substance; and thought constitutes the nature of thinking substance. For everything

[23] The French text is quite different here: "But when it comes to knowing whether one of these substances truly exists, that is, whether it is at present in the world, the fact that it exists without the aid of any created thing is not sufficient to cause us to perceive it; for that fact alone does not reveal to us anything which excites some specific knowledge in our mind."

else which can be attributed to body presupposes extension, and is only a
certain mode {or dependence} of an extended thing; and similarly, all the
properties which we find in mind are only diverse modes of thinking. Thus,
for example, figure cannot be understood except in an extended thing, nor
can motion, except in an extended space; nor can imagination, sensation,
or will, except in a thinking substance. But on the contrary, extension can
be understood without figure or motion; and thought without imagination
or sensation, and so on; as is obvious to anyone who pays attention to these
things.

54.   How we can have clear and distinct notions of thinking and
             corporeal substance, and, similarly, of God.

And thus we can easily have two clear and distinct notions, or ideas; one
of created thinking substance, the other of corporeal substance, provided
of course that we carefully distinguish all attributes of thought from the
attributes of extension. So too we can have a clear and distinct idea of an
uncreated and independent thinking substance, that is, of God: provided
that we do not suppose that this idea adequately represents all things which
are in God, and do not allow our imagination to add anything to it, but
notice only those things which are truly contained in it and which we plainly
perceive to pertain to the nature of a supremely perfect being. And certainly
no one can deny that such an idea of God is in us, unless he judges that there
is absolutely no knowledge of God in human minds.

55.   How duration, order, and number are also distinctly
             understood.

Duration, order, and *number* will also be very distinctly understood by us
if we do not inappropriately attribute any concept of substance to them,
but think that duration is, in each thing, only a mode under which we
conceive of that thing as long as it continues to exist. And similarly, we
must not consider that order is anything diverse from things which are
ordered, or number from things which are numbered, but that they are only
modes under which we consider these things.

56.   What modes, qualities and attributes are.

And indeed here we are understanding by *modes*, exactly the same thing as we understand elsewhere by *attributes* or *qualities*. But when we consider that the substance is affected or altered by these things, we call them *modes*; when the kind of this substance can be named from this alteration, we call them *qualities*; and finally, when we more generally consider these only as being inherent in a substance, {without considering them otherwise than as dependences of that substance}, we call them *attributes*. And therefore, properly, we do not say that there are modes or qualities in God, but only attributes. because no variation is to be understood in Him. And even in created things, those properties which never occur in them diversely, for example, existence and duration in existing and enduring things, ought not to be called qualities or modes, but attributes.[24]

57.   That certain attributes are in things, and others in the mind. And what duration and time are.

However, some attributes or modes are in the things themselves, while others are only in our minds. Thus, when we distinguish time from duration taken in general and say that time is the measure of motion, this is only a mode of thinking; for we certainly do not understand a duration in motion which differs from duration in things which are not moved. This is apparent from the fact that if two bodies are moved for an hour, one slowly and one more rapidly, we shall not count more time in one than in the other, even if there is much more motion. But in order to measure the duration of all things, we compare it with the duration of those greatest and most uniform motions[25] from which years and days are created; and we call this duration "time": which, accordingly, adds nothing to duration taken in general except a mode of thought.

[24] The distinction here seems to be that modes are properties which can vary without altering the nature of a thing in any important way; qualities are those properties whose alteration would produce an alteration in the nature of the thing involved; and attributes are those properties without which the thing cannot exist at all. In a letter written in 1645 or 1646, Descartes states: ". . . I distinguish between Modes properly speaking, and Attributes without which the thing possessing these attributes cannot exist; or between the modes of the things themselves and modes of thinking.... Thus, figure and motion are properly speaking modes of corporeal substance, because the same body can exist, at times with one figure and at times with another, . . . whereas existence, duration magnitude, number, and all universals do not seem to me to be properly called modes.... But they are called by the wider name of Attributes, or modes of thinking,...": A. & T., IV, 348–350; see also Articles 60–62.
[25] That is, the apparent motions of the sun and the fixed stars.

58.   That number and all universals are only modes of thinking.

So also, when number is not being considered in any created things, but only in the abstract or in general, it is merely a mode of thought; as are all the other things which we call *universals*.

59.   How universals are created; and what the five generally known ones; genus, species, difference, property, and accident, are.

These universals are created solely by the fact that we use one and the same idea in order to think of all individual things which are similar to one another: and as we also give one and the same name to all things represented by this idea; this name {also} is universal. Thus, when we see two rocks, and do not pay attention to their nature but only to the fact that there are two of them, we form the idea of their number, which we call binary; and when we afterwards see two birds, or two trees, and still do not consider their nature, but only the fact that there are two of them, we repeat the same idea as before, which accordingly is universal; so that we call this number by the same universal name: 'binary'. Similarly, when we consider a figure bounded by three lines, we form a certain idea of it, which we call the idea of a triangle; and we afterwards use the same idea as a universal, in order to represent in our mind all other figures bounded by three lines. And when we notice that, among triangles, some have one right angle and others have none, we form the universal idea of a right-angled triangle, which, being related to the preceding idea as to a more general one, is called the *species*. And the rightness of that angle is a universal *difference*, by which all right-angled triangles are distinguished from others. And the fact that, in these right-angled triangles, the square of the base[26] is equal to the squares of the sides is a *property* of all these triangles, and only of these. And finally, if we suppose some triangles of this kind to be moved, while others are not moved, this movement will be a universal *accident* in them. And thus five universals are usually counted: *genus, species, difference, property*, and *accident*.

60.   Concerning distinctions, and first, concerning real distinction.

Further, number, in things themselves, arises from their distinction: and

---

[26] Literally: "the power of the base"; the base of a right triangle is its longest side, i.e., the hypotenuse.

this *distinction* is threefold: *real, modal,* and *rational. Real* [*distinction*] properly exists only between two or more substances: and we perceive these to be really distinct from one another from the sole fact that we can clearly and distinctly understand one without the other. For, knowing God, we are certain that He can accomplish whatever we distinctly understand. For example, from the sole fact that we now have the idea of an extended or corporeal substance (although we do not yet know with certainty that any such substance truly exists), we are however certain that it can exist; and that if it exists, each part of it {which can be} delimited by our mind is really distinct from the other parts of the same substance. And, similarly, solely because each of us understands himself to be a thinking thing, and can, by thought, exclude from himself every other substance, both thinking and extended; it is certain that each one, regarded in this way, is really distinguished from every other thinking substance and from every other corporeal substance. And even if we suppose that God has joined some corporeal substance to some such thinking substance so closely that they cannot be more closely joined, and has thus created some one thing from these two; they nonetheless remain really distinct: because, however closely He may have joined them, He cannot have divested Himself of the power which He previously had to separate them, or to conserve one without the other; and things which God can either separate or else conserve separately, are really distinct.

61.   Concerning modal distinction.

Modal distinction is twofold: that is to say, there is one between what is properly called the mode, and the substance of which it is a mode; and another between two modes of the same substance. The first is known from the fact that we can indeed clearly perceive a substance without the mode which we say differs from it, but cannot, conversely, understand the mode without the substance itself. And as figure and motion are modally distinguished from the corporeal substance in which they are; so also affirmation and recollection are modally distinguished from the mind. On the other hand, the second, {which is between two different modes of the same substance}, is known from the fact that we can recognize one mode without the other and *vice versa*; but can recognize neither without that substance to which they belong. So, if a rock is moved and is square, I can indeed understand its square figure without movement; and, conversely, its movement without square figure; but I can understand neither that movement nor that figure without the substance of the rock. However, it seems that the distinction by which the mode of one substance differs from another substance or from the mode of another substance, as [for instance] the movement of one body from another body or from the mind, and as motion from duration, should be called "real", rather than "modal": because those modes are not clearly understood without the really distinguished substances of which they are the modes.

62.   Concerning rational distinction.

Finally, *rational distinction* is between substance and something attributed to it without which the substance itself cannot be understood, or between two such attributes of some single substance. And rational distinction is recognized from the fact that we cannot form a clear and distinct idea of this substance if we exclude that attribute from it; or cannot clearly perceive the idea of one attribute of this kind if we separate it from the other. So, because any substance also ceases to be if it ceases to endure, substance is distinguished from its duration only in the reason. And all modes of thought[27] which we consider as being in objects differ only in the reason; both from the objects of which they are thought, and from one another in one and the same object.[28] I do remember that I elsewhere joined this kind of distinction with modal distinction, specifically, toward the end of the responses to the first objections to the *Meditations on First Philosophy*: but in that place, there was no occasion to differentiate these precisely, and it sufficed for my purpose to distinguish both from real distinction.

63.   How thought and extension can be distinctly known, as constituting the nature of the mind and the body.

Thought and extension can be regarded as constituting the natures of thinking and corporeal substance; and then they must not be conceived

---

[27] That is, universals.

[28] In the letter quoted in note 24, Descartes goes on to say: "Thus, when I think of the essence of a triangle and of the existence of the same triangle; these two thoughts, even taken objectively, differ modally insofar as they are thoughts, and if we take the term 'mode' strictly. But it is not the same with a triangle existing outside the mind, in which it seems obvious to me that existence and essence are in no way distinguished; and the same is true of all universals. For example, when I say, Peter is a man, the thought by which I think of Peter differs modally from the thought by which I think of man; but in Peter himself, to be a man is nothing other than to be Peter, etc."

otherwise than as thinking and extended substance themselves, that is, otherwise than as mind and body; in which way they are most clearly and distinctly understood. Indeed, we more easily understand extended substance or thinking substance than substance alone, when that which thinks or is extended has been omitted. For there is some difficulty in separating the notion of substance from the notions of thought or extension, which of course differ from substance only in the reason; {in that we sometimes consider thought or extension without reflecting on the very thing which thinks or is extended}. And a concept does not become more distinct because we include fewer things in it, but only because we carefully distinguish those things which we do include in it from all others.

64.   How thought and extension can also be distinctly known as modes of substance.

Thought and extension can also be taken as modes of substance, insofar, of course, as one and the same mind can have many diverse thoughts; and as one and the same body can be extended in many diverse ways while retaining its same quantity: that is to say, more in length and less in width or depth at one moment, and, on the contrary, more in width and less in length a little later. And thought and extension are then modally distinguished from substance, and can be understood no less clearly and distinctly than substance itself[29] (provided that they are not regarded as substances, or as certain things separated from others, but only as modes of things). For we distinguish thought and extension from these substances by the fact that we consider them as modes of those substances in which they are; and thus we know what they truly are. And, on the other hand, if we attempted to consider thought and extension without the substances in which they are, by that sole fact we would be regarding them as subsisting things; and thus would be confusing the ideas of mode and substance.

65.   How their modes are also known.

In the same way, we shall best understand the diverse modes of thought, such as understanding, imagining, remembering, willing, etc., and also the diverse modes of extension or those pertaining to extension, such as all

---

[29] The French text says: ". . . and insofar as we do not distinguish thought and extension from that which thinks or is extended, except as we distinguish dependences of a thing from the thing itself; we know them as clearly and as distinctly as their substances, ..."

figures, and situation and movements of parts; if we regard them only as
modes of the things in which they are. And as for motion, we shall best
understand it if we think only of local motion and do not enquire into the
force by which it is produced; although I shall attempt to explain this force
in the proper place.[30]

66.   How sensations, emotions, and appetites are clearly known,
               although we often make incorrect judgments concerning them.

There remain sensations, emotions, and appetites, which can also be
clearly perceived if we carefully avoid making any judgment about them
beyond what is exactly contained in our perception, and what we are
inwardly conscious of. But it is most difficult to observe this, at least where
sensation is concerned: because there is no one among us who did not
judge, at the beginning of his life, that all the things which he observed were
certain things which existed outside his mind and were exactly similar to his
sensations, that is, to the perceptions which he had of them. So that, upon
seeing a color, for example; we thought that we were seeing a certain thing
which was located outside us and exactly similar to the idea of that color
which we were then experiencing in ourselves; and on account of our habit
of thus judging, it seemed to us that we were seeing that so clearly and
distinctly that we held it to be certain and indubitable: {and thus it must not
be thought strange that some men subsequently remain so convinced of this
false and hasty judgment that they cannot bring themselves to doubt it}.

67.   That we often err even in the judgment of pain.

And exactly the same is true of all the other sensations which are felt,
even pleasure and pain. For although these are not thought to be outside
us; they are however not usually regarded as being solely in our mind or
perception, but as being in our hand, or our foot, or some other part of our
body. And it is definitely as uncertain that a pain which we feel as if in the
foot, say, is something existing outside our mind, in the foot; as [it is
uncertain] that the light which we see as if in the Sun exists outside us, in the
Sun {in the way it is in us}: but both these prejudices belong to our
childhood, as will presently be clear.

[30] See Part II, Articles 43 and 44.

68.   How, in these matters, that which we clearly know must be distinguished from that in which we can be deceived.

However, in order that we may distinguish here what is clear from what is obscure, we must most carefully notice that pain, and color, and the remaining things of this kind, are clearly and distinctly perceived when regarded as only sensations or thoughts. However, it must also be noticed that when they are judged to be certain things existing outside our mind, it is absolutely impossible to understand in any way what things they are; and that when someone says that he sees color in some body, or feels pain in some limb, it is exactly as if he were to say that he sees or feels there something of whose nature he is completely ignorant, that is, that he does not know what he is seeing or feeling. For although, while paying insufficient attention, he may easily convince himself that he has some knowledge of it from the fact that he supposes that there is something similar to the sensation of that color or pain which he is experiencing in himself; if however he examines what it is that this sensation of color or pain (considered as if existing in the colored body, or in the painful part) represents {to him}, he will certainly notice that he is entirely ignorant of it.

69.   That size, figure, etc., are known in a very different manner from colors, pains, etc.

[He will] especially [notice this] if he considers that size, in a body which has been observed, or figure, or motion (at least local motion: for Philosophers have rendered the nature of motion less intelligible to themselves by imagining certain other different sorts of motion), or situation, or duration, or number, and other similar things which we have already stated are perceived clearly in bodies; are known by him in a manner very unlike that in which he knows, in the same body, what color is, or pain, or odor, or flavor, or any of the other things which I have said must be referred to the senses. For when we observe some body, although we are as certain that it exists insofar as it appears to have color as we are insofar as it appears to have figure; yet we know much more clearly what it is for that body to have figure than what it is for it to have color.

70.   That we can make a judgment concerning perceptible things in two ways; in one of which we avoid error, while in the other we fall into error.

And so it is obvious that when we say that we perceive colors in objects, this is in fact the same as if we were to say that we perceive something in objects of whose nature we are ignorant, but by means of which a certain very manifest and evident sensation is created in us, which is called the sensation of colors. However, there is a very great difference in the ways of judging [associated with these two remarks]: for as long as we merely judge that there is something in objects (that is, in the things from which a sensation comes to us, of whatever exact kind those things may be) the nature of which we do not know; we will be so far from being deceived that we will instead avoid error; because when we notice that we are ignorant of something, we are less inclined to judge rashly of it. But when we think that we perceive colors in objects, although in fact we do not know what it is that we are then calling by the name 'color', and cannot understand any similarity between the color which we are supposing to be in objects and that which we experience to be in our sensation; however, because we do not notice this very fact (and because there are many other things, like size, figure, number, etc., which we clearly perceive to be felt or understood by us in a manner which does not differ from that in which they are, or at least can be, in objects): we easily fall into the error of judging that that which we call color in objects, is something entirely similar to our sensation of color; and thus of believing we clearly perceive something which [in fact] we do not perceive in any way.

71.   That the principal cause of errors proceeds from the prejudices of our childhood.

And here the first and principal cause of all errors can be recognized. For in childhood, our mind was of course so closely bound to the body that it did not apply itself to any thoughts other than those by means of which it was aware of those things which affected the body: and it did not yet relate those to something situated outside itself; but merely felt pain when something disagreeable occurred to the body; and pleasure when something agreeable occurred. And when the body was affected without great advantage or disadvantage, according to the diversity of the parts of and the ways in which the body was affected,[31] the mind had certain diverse sensations, namely those which we call the sensations of taste, of odor, of sound, of heat, of cold, of light, of colors, and of similar things; which represent nothing situated outside thought. And at the same time, the mind

[31] The French text reads: "... according to the diversities occurring in the movements which travel from all parts of our body to the place in the brain to which the soul is so closely joined and united."

also perceived sizes, figures, motions, and such; which were not presented to it as sensations, but as certain things or modes of things, existing or at least capable of existing, outside thought: even if it did not yet note this [latter] distinction between those things. And next, when the mechanism of the body (which was made by nature in such a way that it can be moved by its own power in various ways) turned itself randomly this way and that and happened to pursue something pleasant or to flee from something disagreeable; the mind attached to it began to notice that that thing which the body thus pursued or avoided was outside itself, and did not only attribute to it sizes, figures, motions and such (which it perceived {very clearly} as things or as modes or things), but also [attributed to it] flavors, odors, and the rest; the sensations of which the mind noticed were produced in it by that thing. And relating all things solely to the utility of the body in which it was immersed, the mind thought that there was more or less substance in each object which affected the body, accordingly as the body was more or less affected by that object. As a result, the mind thought that there was much more substance or corporeality in rocks or metals than in water or air; because it perceived more hardness and weight in the former. Indeed it esteemed the air as absolutely nothing, as long as it experienced in it no wind or heat or cold. And because no more light shone upon it from the stars than from the tiny flames of lamps; it accordingly represented stars to itself as being no larger than those flames. And because it did not note that the earth was rotated or that its surface was curved like a globe, it was therefore more inclined to think both that it was immobile and that its surface was flat. And our mind has been filled from earliest childhood with a thousand other prejudices of this kind; which it subsequently, in youth, did not remember having adopted without sufficient examination but accepted as most true and evident; as if known by perception or imparted to it by nature.

72.   That the second cause of error is that we are unable to forget our prejudices.

And although now, in mature years, our mind is no longer entirely subordinated to the body and no longer relates everything to it but also enquires about the truth of things considered *per se*; it perceives that very many of those judgments which it thus formerly made are false. However, that does not make it easy for the mind to erase those judgments from its memory; and as long as they remain in it, they can be the causes of various errors. Thus, for example, because in our earliest years we imagined very tiny stars (even though Astronomical reasonings now show us clearly that the stars are extremely large), our prematurely formed opinion is nonetheless still so strong that it is most difficult for us to imagine them otherwise than before.

73.   That the third cause is that we grow tired through concentrating upon those things which are not present to the senses; and consequently are not accustomed to judge them from present perception, but from preconceived opinion.

Besides, our mind cannot concentrate upon any thing without some difficulty and fatigue; and it concentrates with the greatest difficulty of all upon those things which are not present either to the senses or even to the imagination: either because such is the nature of the mind (inasmuch as it is joined to the body); or because in earliest years, when it was only concerned with sensations and imaginings, it acquired greater practice and facility in thinking about these than about the remaining things. Moreover, as a result of this, many men now comprehend no [kind of] substance, except imaginable, corporeal, and actually perceptible. For they do not know that only those things which are characterized by extension, motion, and figure are imaginable, even though many others may be intelligible; and they do not think that anything can subsist which is not a body; or, finally, that any body which is not perceptible [can subsist]. And because in fact we cannot perceive anything as it [truly] is solely by means of the senses, as will be clearly shown later; the result is that most men never perceive anything in their whole lives except in a confused way.

74.   That the fourth cause is that we attach our concepts to words which do not accurately correspond to things.

And finally, because of the use of speech, we attach all our concepts to words by which we express them, and do not commit them to memory except along with these words. And since we afterwards more easily remember the words than the things; we scarcely ever have a concept of any thing so distinct that we separate it from all conception of the words {chosen to express it}; and the thoughts of almost all men are occupied with words more than with things. Thus men very often give their assent to words which they have not understood; because they think that they formerly understood them or learned them from others who understood them correctly. And although these things cannot be propounded here with exactitude, because the nature of the human body has not yet been explained, and because it has not yet been proved that any bodies exist; it nonetheless seems that they can be understood sufficiently to assist in distinguishing clear and distinct concepts from obscure and confused ones.

75. A summary of those things which must be observed in order to philosophize correctly.

And so, in order to philosophize seriously, and to discover the truth about all things which can be known: first, all prejudices must be abandoned or else we must carefully avoid trusting any of the opinions accepted by us in the past unless we first ascertain that they are true by submitting them to a new examination. Next, in order to proceed correctly, we must pay attention to the notions which we ourselves have in us; and all those, and only those, which we clearly and distinctly know while we thus attend to them, are to be judged true. In doing so, we shall first notice that we exist, insofar as our nature is that of a thinking thing; and at the same time we shall also notice both that God is, and that we depend upon Him, and that from a consideration of His attributes we can investigate the truth of the remaining things, since He is their cause. Finally, we must note that, in addition to the notions of God and of our mind, there is also in us the knowledge of many statements of eternal truth, for example, that no thing can be produced from nothingness, etc.; and similarly, there is the knowledge of a certain corporeal nature, or one extended, divisible, mobile, etc.; and also the knowledge of certain sensations which affect us, for example, pain, colors, flavors, etc. (although we do not yet know what causes us to be thus affected). And comparing these things with those which we formerly more confusedly thought, we shall acquire the practice of forming clear and distinct conceptions of all things which can be known. And the chief principles of human knowledge seem to me to be contained in these few.

76. That divine authority is to be preferred to our perception: but that, apart from divine authority, it does not become a philosopher to assent to things other than those which have been perceived.

However, in addition to the rest, it must be firmly fixed in our memory as the supreme {infallible} rule that those things which have been revealed to us by God must be believed to be the most certain of all. And that, although perhaps {some spark of} the light of reason might seem to very clearly and evidently suggest to us something else; nonetheless, trust must be placed solely in divine authority rather than in our own judgment. But in those matters about which divine faith teaches us nothing, it by no means becomes a philosopher to accept as true something which he has never perceived to be true; and to trust in his senses, that is, the unconsidered judgments of his childhood, more than in mature reason.

PART II

OF THE PRINCIPLES OF
MATERIAL OBJECTS

PART II

1.   The reasons why we know with certainty that material objects exist.

Even though there is no one who is not sufficiently convinced that material objects exist, nonetheless, because we called this issue into question a short while ago and included it among the prejudices of our early youth, we must now investigate the reasons why this may be known with certainty. Of course, {we experience in ourselves that} whatever we feel undoubtedly comes to us from something different from our mind. For it is not in our power to cause ourselves to feel one sensation rather than another; on the contrary, this plainly depends on whatever is influencing our senses. We can of course ask whether this may be God, or something different from God. But because we feel, or rather, because our senses lead us to clearly and distinctly perceive, a certain matter which is extended in length, breadth, and depth (the diverse parts of which are endowed with various shapes and subject to diverse movements); and which also causes us to have sensations of color, odor, pain, etc. : if God were Himself directly presenting the idea of this extended matter to our mind, or even merely causing it to be presented by something which lacked extension, shape, and movement; it would be impossible to devise any reason for not thinking Him a deceiver. For we clearly understand this supposed thing to be completely distinct, not only from God, but also from us or from our mind. Moreover, we seem to see clearly that the idea of it comes from external things, which it perfectly represents; and, of course, as has already been noticed, it is completely contrary to God's nature to be a deceiver.[1] It must therefore be concluded with certainty that there exists a certain substance, extended in length, breadth, and depth, and possessing all those properties

---

[1] This argument is a bit more complex than it appears. The fact that our sensations incline us to believe that there are material objects whose nature is as we clearly and distinctly understand it to be does not in itself prove that material objects exist, or that God would be deceiving us if they did not. Rather, it is the fact that we cannot verify or refute this belief by the use of our reason that would make it a deception if it were untrue. See Part I, Article 60. Many of the beliefs which arise from our senses are false, on Descartes's view; but they can be known to be false: cf. Articles 3 and 4.

which we clearly conceive to be appropriate to extended things; and it is this extended substance which we call body or matter.

2.   And the reasons why we know that the human body is united
                with the mind.

In the same way, it can be concluded that a particular body is united with our mind more closely than any other bodies are. This is obvious from the fact that we clearly notice that pains and other sensations come to us unexpectedly, and that our mind is conscious that these sensations do not proceed from it alone, and cannot pertain to it solely in virtue of its being a thinking thing. Rather, they proceed from it only because it is joined to some other thing which has extension and is mobile, and which is called the human body. However, this is not the place for a more precise explanation of this thing.

3.   That the perceptions of our senses do not teach what really
                exists in things, but only what can harm or benefit that union.

It will suffice for us to notice that the perceptions of our senses pertain only to this union of a human body with a mind, and that, even though they generally show us how external bodies can be beneficial or harmful to this union, they do not, however (except occasionally and accidentally), teach us what these things are like in themselves. We shall thereby easily lay aside those prejudices which arise {solely} from our senses, and shall use here only our understanding; by carefully concentrating it on those ideas with which nature endowed it, {and which are like the seeds of those truths which we are capable of knowing}.

4.   That the nature of body does not consist in weight, hardness,
                color, or other similar properties; but in extension alone.

By so doing, we shall perceive that the nature of matter, or of body considered in general, does not consist in the fact that it is hard, heavy, colored, or affects the senses in any other way; but only in the fact that it is a thing possessing extension in length, breadth, and depth. For as far as hardness is concerned, our senses tell us nothing about it except that the parts of hard bodies resist the movement of our hands when they encounter them. Besides, if whenever our hands moved in a certain direction, the bodies situated there were to move back at the speed at which our hands approach; we would {certainly} never feel any hardness. Yet it cannot in any way be understood that the bodies which would thus move back would thereby cease to have the nature of a body. Therefore, the nature of body does not consist in hardness. In the same way, it can be shown that weight, color, and all the other properties of this kind which are experienced in material substance, can be taken away; leaving that substance intact. From this it follows that the nature of matter does not depend on any such properties, {but consists solely in the fact that it is a substance which has extension}.[2]

5.   That prejudices concerning rarefaction and the void obscure [this truth about] the real nature of body.

However, there still remain two causes which might lead one to doubt whether the true nature of body consists in extension alone. The first is the common belief that many bodies can be rarefied and condensed in such a way as to have more extension when rarefied than when condensed; and there are even some men so subtle that they distinguish the substance of a body from its quantity {or size}, and this quantity from its extension. The other cause is that we are not accustomed to say that there is a body in those places where we understand that there is nothing other than extension in length, breadth, and depth; rather, we say that there is only space, and moreover, empty space; which almost everyone believes to be complete nothingness.

6.   How rarefaction occurs.

As far as rarefaction and condensation are concerned, whoever thinks carefully and resolves to accept only what he clearly {and distinctly} perceives, will believe that nothing other than change of shape is involved in these events. Thus, rarefied bodies are those with many spaces between their parts which are filled by other bodies. And rarefied bodies only become denser when their parts, by approaching one another, either

---

[2] An important consideration, which is not made explicit here, is that *only* extension, figure, and motion are capable of being clearly and distinctly perceived by the understanding. They are also the only properties which can be directly represented geometrically; see Part I, Article 69. Consequently, only those properties are capable of generating necessary truths about bodies.

diminish or completely eliminate these spaces; if the latter ever occurs, then the body grows so dense that it cannot possibly become denser. However, it does not then have less extension than it did when it filled a greater space because its parts were separated from one another. For whatever extension there is in the spaces between its parts must in no way be attributed to it, but to whatever other bodies fill those spaces. Thus, when we see a sponge full of water or another liquid, we do not think that, in terms of its own individual parts, it has more extension than when it is compressed and dry; but only that its pores are more open, and that its parts are therefore spread over more space.

7.　　　　That rarefaction cannot be intelligibly explained in any other way.

And certainly I do not see why some men prefer to say that rarefaction occurs by increase in quantity, rather than to explain it by this example of the sponge. For although when air or water becomes rarefied we do not see its pores becoming larger, or any new body approaching to fill them; it is less consistent with reason to imagine something unintelligible, in order to {appear to} explain rarefaction by a merely verbal device, than it is to conclude, from the fact that bodies become rarefied, that they contain pores or interstices which grow larger and that some new body approaches to fill these pores; even though we may not perceive this new body through any of our senses. For there is no reason why we should believe that all bodies which exist must affect our senses. Besides, we perceive that rarefaction can very easily occur {and be explained} in this way, though in no other. Finally, it is clearly contradictory for anything to be increased by new quantity or new extension, unless a new extended substance, that is, a body, is added to it: for it cannot be understood that any increase in extension or quantity can occur except by the addition of a substance which has extension and quantity, as will be made even clearer by what follows.

8.　　　　That quantity[3] and number differ from the thing which has quantity and is numbered only in our manner of conceiving them.

For quantity does not in fact differ from the extended substance except insofar as our conception of it is concerned; similarly, number does not

[3] By 'quantity' here Descartes means 'volume' or 'extension'; see Article 9.

differ from the thing which is numbered. Thus, we can consider the essential nature of a corporeal substance which occupies a volume of ten feet even though we may pay no attention to this measurement of ten feet; because the body's nature is understood to be exactly the same in any part of this volume as in the whole. And *vice versa*, the number ten, or even a volume of ten feet, can be comprehended even if we pay no attention to the particular substance in that space; for the conception of the number ten is exactly the same whether it concerns this measurement of ten feet or something else. And a volume of ten feet, although it cannot be comprehended without [the notion of] some extended substance, of which it is the quantity, can however be comprehended apart from any particular body. Yet in fact it cannot happen that the least part is taken away from this quantity or extension, without an equal amount of substance being removed. Nor, *vice versa*, can it occur that even a trifle is removed from the substance, without an equal amount of quantity or extension being taken away.

9.   That corporeal substance, when distinguished from its quantity {or extension}, is confusedly conceived as if it were incorporeal.

And although others may perhaps say something else, I do not however think that they have a different perception of this question. For when they distinguish substance from extension, or quantity; either they understand nothing by the word 'substance', or they have a confused idea of some sort of incorporeal substance, whose nature they falsely attribute to corporeal. And they call the true idea of corporeal substance 'extension', which they, however, call an accident; and thus they proclaim in words something quite different from what they themselves comprehend in their minds.

10.   The nature of space or internal place.[4]

Nor in fact does space, or internal place, differ from the corporeal substance contained in it, except in the way in which we are accustomed to conceive of them. For in fact the extension in length, breadth, and depth which constitutes the space occupied by a body, is exactly the same as that which constitutes the body. The difference consists in the fact that, in the body, we consider its extension as if it were an individual thing, and think that it is always changed whenever the body changes. However, we

---

[4] By the "internal place" of an object Descartes means the volume the body occupies. External place is, roughly, its location with respect to other bodies.

attribute a generic unity to the extension of the space, so that when the body
which fills the space has been changed, the extension of the space itself is not
considered to have been changed {or transported} but to remain one and
the same; as long as it remains of the same size and shape and maintains the
same situation among certain external bodies by means of which we specify
that space.

11.   That space does not in fact differ from material substance.

Further, if we concentrate on the idea which we have of some body, for
example a stone, and remove from that idea everything which we know is
not essential to the nature of body; we shall easily understand that the same
extension which constitutes the nature of body also constitutes the nature
of space, and that these two things differ only in the way that the nature of
the genus or species differs from that of the individual. We may certainly
begin by removing hardness; for if the stone melts or is reduced to the finest
possible powder, it will lose hardness but will not thereby cease to be a
body. We may also remove color, for we have often seen stones so
transparent that they had no color; we may take away weight, because,
although fire is extremely light, it is nonetheless thought to be a body.
Finally, we may take away cold, heat, and all other properties which are
either not considered to be always in the stone, or which could be changed
without the stone being thought to have lost the nature of body. For then
we shall clearly notice that absolutely nothing remains in our idea of the
stone except that {we distinctly perceive that} it is something extended in
length, breadth, and depth; and this fact is also included in our idea of
space, and not only of space which is full of bodies, but also of space which
is called a void.[5]

12.   How space differs from material substance in the way in which
             it is conceived.

There is however a difference in our way of conceiving them; for when a
stone has been removed from the space or place in which it was, we think
that its extension has also been removed; since we regard that extension as
unique and inseparable from the body. However, we judge that the
extension of the place in which the stone was remains and is the same,

[5] That is, material substance and space have the same essential nature and thus are one and the
same substance. See Article 16.

although the stone's place may now be occupied by wood, or water, or air, or any other body; or may even be believed to be empty. The reason for this is that extension in that case is being considered in a general way; and thus the same extension can be thought to be common to stone, wood, water, air, or other bodies, or even to a vacuum itself (if any is assumed to exist) provided only that it is of the same size and shape and maintains the same situation among the external bodies which determine that space.

13.   What external place is.

For in fact the names 'place' or 'space' do not signify a thing different from the body which is said to be in the place; but only designate its size, shape, and situation among other bodies. Moreover, in order to determine that situation we must take into account some other bodies which we consider to be motionless: and, depending on which bodies we consider, we can say that the same thing simultaneously changes and does not change its place. Thus, when a ship is heading out to sea, a person seated in the stern always remains in one place as far as the parts of the ship are concerned, for he maintains the same situation in relation to them. But this same person is constantly changing his place as far as the shores are concerned, since he is constantly moving away from some and toward others. Furthermore, if we think that the earth moves {and is rotating on its axis}, and travels from the West toward the East exactly as far as the ship progresses from the East toward the West; we shall once again say that the person seated in the stern does not change his place: because of course we shall determine his place by certain supposedly motionless points in the heavens. Finally, if we think that no truly motionless points of this kind are found in the universe, as will later be shown to be probable;[6] then, from that, we shall conclude that nothing has an enduring {fixed and determinate} place, except insofar as its place is determined in our minds.

14.   The respects in which place and space differ.

However, the names 'place' and 'space' differ, because 'place' designates situation more specifically than extension or shape; and, on the other hand, we think more specifically of the latter when we speak of space. For we

---

[6] The French text reads, ". . . if we think that it is impossible to discover any point in the entire universe which is truly motionless (and the reader will learn from what follows that this can be proved). . ."

frequently say that one thing takes the place of another although it is not of precisely the same size or shape; but then we are [implicitly] denying that it occupies the same space as the other did. Further, when a body's situation changes, we say that its place changes, although the same size and shape may remain. So when we say that a thing is in a certain place, we understand only that it is in a certain situation in relation to other things; but when we add that it fills that space, or that place, we understand also that it has the specific size and shape of that space.

15.   How external space is correctly taken to be the surface of the surrounding body.[7]

Thus, we always take space to be extension in length, breadth, and depth. However, we sometimes consider the place of a thing as its internal place, {as if it were in the thing placed}; and sometimes as its external place, {as if it were outside this thing}. In fact, internal place is exactly the same as space; while external place can be taken to be the surface which most closely surrounds the thing placed. It must be noticed that by 'surface' we do not understand here any part of the surrounding body, but only the boundary between the surrounding and surrounded bodies, which is simply a mode. Or to put it another way, we understand by 'surface' the common surface, which is not a part of one body more than of the other, and which is thought to be always the same provided that it retains the same size and shape. For even if the whole surrounding body, with its surface, is changed; we do not on that account judge that the surrounded thing changes its place if it maintains the same situation among those external bodies which we consider to be at rest. For example, if we suppose a boat to be driven in one direction by the flow of a river, and in the other by the wind, with perfectly equal force (so that it does not change its situation between the banks), anyone will easily believe that it remains in the same place, although all its surrounding surfaces change.

16.   That it is contradictory for a vacuum,[8] or a space in which there is absolutely nothing, to exist.

That a vacuum in the philosophical sense of the term (that is, a space in

[7] This is Aristotle's definition of 'place'. Part of Descartes's purpose in these passages is to deny the Aristotelian view that place or space is something distinct from body. The term translated as 'surface' is 'superficies'.

[8] The Latin term is 'vacuum', which means 'void', 'vacuum', or 'emptiness'.

which there is absolutely no [material] substance) cannot exist is evident from the fact that the extension of space, or of internal place, does not differ from the extension of body. From the sole fact that a body is extended in length, breadth, and depth; we rightly conclude that it is a substance: because it is entirely contradictory for that which is nothing to possess extension. And the same must also be concluded about space which is said to be empty: that, since it certainly has extension, there must necessarily also be substance in it.[9]

17.   That the word 'void', in common usage, does not exclude all body.

Indeed, in common usage, we do not usually mean, by the word 'void', {or 'empty'}, a place or space in which there is absolutely nothing, but only a place in which there are none of those things which we think ought to be in it. Thus, because an urn is made to contain water, it is said to be empty when it is only filled with air. Thus too, if there are no fish in a fish-pond, it is empty, even though full of water; again, a ship which usually carries merchandise is empty if loaded only with sand to serve as ballast. And, in the same way, a space is said to be void if it contains nothing perceptible, even though it may be full of created matter subsisting by itself;[10] because we are accustomed to consider only those things which are perceived by our senses. If henceforth, instead of heeding what ought to be understood by the words 'void' and 'nothing', we think that a space which we have stated to be empty contains, not merely nothing perceptible, but absolutely nothing at all; we shall be making the same mistake as if, from the fact that it is customary to say that an urn in which there is nothing but air is empty, we were to judge that the air contained in it is not a substance.

18.   How our prejudice concerning [the possibility of] an absolute vacuum must be corrected.

Almost all of us made this mistake from the beginning of our lives, because (not observing any necessary connection between a vessel and the

[9] That is, since extension is a property, it must be a property of some substance. See Part I, Article 11.

[10] In a letter to Hyperaspistes written in August, 1641, Descartes says, ". . . when we say of created substance that it subsists by itself we are not thereby excluding that divine participation which it needs to subsist; but we only mean that it . . . can exist without any other created thing . . .": A. & T., III, 429.

particular body contained in it) we thought that there was nothing to prevent at least God from causing the body which fills some vessel to be removed without any other taking its place. Now in order to correct this error, we must consider that, although there is no connection between the vessel and whatever particular body is contained in it, there is a very great and absolutely necessary connection between the concave shape of the vessel and the extension, taken in a general sense, which must be contained in that concavity. Thus, it would be as contradictory of us to conceive of a mountain without a valley, as to conceive of this concavity without the extension contained in it, or of this extension without an extended substance: because, as has frequently been said, nothingness cannot possess any extension. Accordingly, if anyone asks what would occur if God removed the whole body contained in any vessel and did not permit anything else to take the place of the body which had been removed, the answer will have to be that the sides of the vessel would thereby become contiguous to each other. For, when there is nothing between two bodies, they must necessarily touch each other; and it is manifestly contradictory for them to be apart, or for there to be distance between them, and yet for this distance to be nothing: because all distance is a mode of extension, and therefore cannot exist without an extended substance.

19.   That this confirms what has been said concerning rarefaction.

After thus observing that the nature of a material substance consists solely in the fact that it is an extended thing; and that its extension does not differ from that which is usually attributed to a perfectly empty space; we shall easily recognize that it is not possible for any one of its parts to occupy more space at one time than at another, nor, consequently, for it to be rarefied in any way other than that explained a little earlier. We shall recognize as well that it is not possible for there to be more matter, or material substance, in a vessel when it is full of lead, gold, or another extremely heavy and hard body, than when it contains only air and is thought to be empty: because the quantity of matter does not depend on the weight or hardness of its parts, but on extension alone, and this is always the same in a given vessel.

20.   That this also shows that no atoms can exist.

We also easily understand that it is not possible for any atoms, or parts of matter which are by their own nature indivisible, to exist. The reason is that if there were any such things, they would necessarily have to be extended, no matter how tiny they are imagined to be. We can, therefore, still conceive of each of them being divided into two or more smaller ones, and thus we know that they are divisible. For it is impossible to {clearly and distinctly} conceive of dividing anything without knowing, from that very fact, that it is divisible; because if we were to judge that same thing to be indivisible, our judgment would be in disagreement with out knowledge [of it]. Moreover, even if we imagine that God wished to create a particle of matter which was impossible to divide into smaller ones; that particle could not, even then, be properly called indivisible. For even supposing that He has made it such that no created being could divide it, He certainly cannot have deprived Himself of His ability to divide it; because, as we noticed earlier, it is absolutely impossible for Him to diminish His own power. Therefore, strictly speaking, this particle will remain divisible, since it is so by virtue of its own nature.

21.   And [this shows], furthermore, that the world is indefinitely extended.

In addition, we understand that this world, or the universe of material substance, has no limits to its extension. For wherever we may imagine those limits to be, we are always able, not merely to imagine other indefinitely extended spaces beyond them; but also to clearly perceive that these are as we conceive them to be, and, consequently, that they contain an indefinitely extended material substance. Because (as has now been shown at length), the idea of that extension which we conceive in any space whatever, is exactly the same as the idea of material substance.

22.   And this shows, similarly, that the matter of the heaven and the earth is one and the same; and that there cannot be a plurality of worlds.[11]

From this it can also be easily inferred that the matter of the heaven does not differ from that of the earth; and that even if there were countless worlds in all, it would be impossible for them not to all be of one and the

---

[11] This is a very important and far-reaching claim. Since all bodies have the same essential nature, they must all obey the same natural laws. This is in sharp contrast to the Aristotelian view that the material of the Earth is different in kind from that of the heavens, and that consequently, different laws of nature apply in each case.

same [kind of] matter. And therefore, there cannot be several worlds, but only one: because we clearly understand that this matter (the nature of which consists solely in the fact that it is an extended substance) now occupies absolutely all the conceivable spaces in which those other worlds would have to be. Nor can we discover, in ourselves, the idea of any other [kind of] matter.

23.   That all the variation in matter, or all the diversity of its forms, depends on motion.

Therefore, all the matter in the whole universe is of one and the same kind; since all matter is identified [as such] solely by the fact that it is extended. Moreover, all the properties which we clearly perceive in it are reducible to the sole fact that it is divisible and its parts movable; and that it is therefore capable of all the dispositions which we perceive can result from the movement of its parts. For although our minds can imagine divisions {in that matter}, this [imagining] alone cannot change matter in any way; rather, all the variation of matter, or all the diversity of its forms, depends on motion. Further, this seems to have been noticed by Philosophers everywhere; because they have said that nature is the principle of motion and rest. And by 'nature', they then understood that by means of which all corporeal things become as we experience them to be.

24.   What movement is in the ordinary sense.

However, movement (and I mean local movement, because I can conceive no other kind, and because I consequently think that no other should be imagined in the nature of things), as commonly interpreted, is nothing other than *the action by which some body travels from one place to another*. And, therefore, in the same way as we have shown above that the same thing can be said to simultaneously change, and not change, its place; so it can also be said to move and not to move. Thus a man, seated in a ship which is sailing out of port, thinks that he is moving if he turns his attention to the shores, which he considers to be at rest. But he does not think so if he turns his attention to the parts of the ship, in relation to which he constantly maintains the same situation. In fact, inasmuch as we commonly think that there is action in all movement, and, on the other hand, cessation of action in rest; he is more properly said to be at rest rather than in motion, because he feels no action in himself, {and because that is customary}.

25. What movement properly speaking is.

If, however, we consider what should be understood by movement, according to the truth of the matter rather than in accordance with common usage (in order to attribute a determinate nature to it): we can say that it is *the transference of one part of matter or of one body, from the vicinity of those bodies immediately contiguous to it and considered as at rest, into the vicinity of [some] others.*[12] By *one body*, or *one part of matter*, I here understand everything which is simultaneously transported; even though this may be composed of many parts which have other movements among themselves. I also say that it is a *transference*, not the force or action which transfers, in order to show that this motion is always in the moving body and not in the thing which moves it (because it is not usual to distinguish between these two with sufficient care); and in order to show that it is only a mode [of the moving body], and not a substance,[13] just as shape is a mode of the thing shaped, and rest, of the thing which is at rest.

26. That no more action is required to produce movement than to bring about its cessation.

Moreover, it must be noticed that we are laboring under a great prejudice when we judge that more action is required for movement than for rest. The reason why we convinced ourselves of this at the beginning of our life is that our body normally moves as the result of a conscious effort of our will, while it remains at rest by the sole fact of being attached to the earth by weight, the force of which we do not feel. Since this weight and many other factors which we do not {usually} notice resist the movements which we seek to produce in our limbs and cause us to become tired, we think that greater action or force is needed to produce movement than to stop it; we

[12] This definition forms the basis of Descartes's view, expressed in Articles 26–31, that our ordinary notion of motion is necessarily relative, and simply consists in the idea of a change in distance between two bodies. Since change in distance is a relational property, Descartes can claim that the property of A, say, which makes us call it a moving body (change in distance from C), is necessarily also a property of C, relative to A. On Descartes's view, we normally select a frame of reference somewhat arbitrarily and then describe motions within that framework as if it were at rest, whereas the proper reference frame is those bodies contiguous to the moving one. This view will later enable Descartes to claim that the Copernican system does not require the motion of the Earth; see note 76, and Part III, Articles 25–30.
[13] Part of Descartes's purpose here is to deny one version of the Medieval theory of impetus. On that view, projectile motion was explained by supposing that some substance, called "impetus", was transferred from the mover to the projectile.

are, of course, [mistaken in] taking action to be the effort required to move our limbs and, by their application, other bodies. However, we shall have no difficulty in correcting this {false} prejudice if we consider that effort is required on our part, not only to move external bodies, but often also to stop their movements, when these are not halted by weight or by another cause[14]. For example, we use no more action to set in motion a boat which is at rest in {calm} water with no current than we use to stop it suddenly while it is moving; and if, {in this case, experience shows us that} a little less action is needed {to stop the boat than to start it moving}, that is because we must take into account the weight and viscosity of the water which the boat pushes aside as it moves and which can gradually bring it to a halt.

27.   That movement and rest are merely diverse modes of the body
                in which they are found.

We are not concerned here with the action which is understood to be in whatever initiates or stops the motion of a body, but only with the body's transference and absence of transference, or rest. It is obvious that this transference cannot exist apart from the body which is moved, and that it is only a case of the body being differently inclined when it is being transported than when it is not transported, or is at rest. Thus, movement and rest are merely two diverse modes of that body.[15]

28.   That movement, properly understood, concerns only the bodies
                contiguous to the body which is moving.

I have also added that the transference is effected *from the vicinity of those bodies contiguous to it into the vicinity of others*, and not from one place to another; because, as has been explained above, 'place' can be understood in several ways, depending on our conception. However, when we take movement to be the transference of a body from the vicinity of

[14] While Descartes frequently states that his physics is deduced from his metaphysics, it is extremely important to realize that his physics is, nonetheless, completely mechanical. That is, an object's motion can be altered only by the impact of another body. In a letter to de Beaune, written in April, 1639, Descartes states: "... all my Physics is merely Mechanics ...''; A. & T., II, 541–544.

[15] Part of Descartes's purpose here (and in Article 25) is to deny the Aristotelian notion that some bodies have a natural tendency toward motion and others a natural tendency toward rest.

those contiguous to it, we cannot attribute to that moving body more than one movement at any given time;[16] because at any given time, only a certain number of bodies can be contiguous to it.

29. And that moreover, it concerns only such of those contiguous bodies as we consider to be at rest.

Finally, I have stated that this transference is effected from the vicinity, not of any contiguous bodies, but only of *those which we consider to be at rest*. For the transference is reciprocal; and we cannot conceive of the body AB being transported from the vicinity of the body CD without also understanding that the body CD is transported from the vicinity of the body AB, and that exactly the same force and action is required for the one transference as for the other. Thus, if we wish to attribute to movement a nature which is absolutely its own, without referring it to any other thing; then when two immediately contiguous bodies are transported, one in one direction and the other in another, and are thereby separated from each other; we should say that there is as much movement in the one as in the other. However, {I admit that} that would depart greatly from the usual manner of speaking; for we are on the earth, and think it to be at rest; and the fact that we see some of its parts, which touch other smaller bodies, being transported from the vicinity of these bodies does not cause us to conclude that the earth is moved.

30. Why the movement which separates two contiguous bodies is attributed to one rather than to the other.

The main reason for this is that we do not think a body moves unless it moves as a whole, and thus we cannot understand that the whole earth moves just because some of its parts are transported from the vicinity of some other smaller bodies which touch them; because we often notice around us many such transferences which are contrary to one another. For example, if the body EFGH is the earth,[17] and if, upon its surface, the body AB is transported from E toward F at the same time as the body CD is transported from H toward G; then even though we know that the parts of the earth contiguous to the body AB are transported from B toward A, and that the action employed in this transference must be neither different in nature nor weaker in the parts of the earth than in the body AB; we do not

[16] For a fuller explanation of this point, see Articles 30 and 31.
[17] See Plate I, Fig. i.

on that account understand that the earth moves from B toward A, or from the East toward the West;[18] because in view of the fact that those of its parts which touch the body CD are being similarly transported from C toward D, we would also have to understand that the earth moves in the opposite direction, i.e., from West to East; and these two statements contradict each other.[19] Accordingly, lest we deviate too far from the customary manner of speaking, we shall say that the bodies AB and CD, and others like them, move; and not the earth. Meanwhile, however, we must remember that all the real and positive properties which are in moving bodies, and by virtue of which we say that they move, are also found in those contiguous to them, even though we consider the second group to be at rest.

31. How there can be innumerable diverse movements in the same body.

Each individual body has only one movement which is peculiar to it, since it is understood to move away from only a certain number of bodies contiguous to it and which are considered at rest; nevertheless, it can also participate in innumerable other movements, inasmuch as it is a part of other bodies which have other movements. For example, if a sailor travelling on board his ship is wearing a watch; although the wheels of his watch will have only a single movement peculiar to them,[20] {it is certain that} they will also participate in that of the voyaging sailor, for they and he together form one body {which is transported as a unit}; they will also participate in the movement of the ship tossing on the ocean, and in that of the ocean itself, {because they follow its currents}; and, finally, in that of the earth, if {one supposes that} the entire earth is moved, {because they form one body with it}. All of these movements will indeed be in the wheels of the watch; but because we do not ordinarily conceive of so many movements at one time, and because we cannot even know all {those in which the wheels of the watch participate}; it will suffice for us to consider in each body the

[18] The text has the terms 'East' and 'West' transposed throughout this article; which is contrary to the illustration.

[19] The contradiction arises only because the two motions are attributed to the Earth absolutely; relative to AB, the Earth does move from East to West, and if CD is considered at rest, then the Earth and AB move from West to East. The French text says: "... ; and there would be too much confusion in this."

[20] That is, a movement relative to the watch-case in which *only* the wheels participate: the movement produced by the watch-spring.

one movement which is peculiar to it {and of which we can have certain knowledge}.

32.   How, properly understood, the single movement peculiar to each body may also be regarded as multiple.

We can even consider this single movement which is peculiar to each body as equivalent to several {separate} movements: thus we distinguish two in the wheels of a carriage, that is to say, one circular, effected around the axle; and the other straight, along the length of the route they take. However, that these two movements are not thereby truly distinct from each other is evident from the fact that each point {of these wheels and} of any {other} moving body describes no more than one line.[21] Nor does it matter that this line is often exceedingly crooked, so that it seems to have been produced by many different movements: for one can imagine any line whatever, even a straight one, which is the simplest of all, to have been described by innumerable diverse movements. For example, if at the same time as the line AB[22] moves toward CD, its point A moves closer to B; the straight line AD (which will be described by the point A), will depend no less on the two movements of A toward B and of AB toward CD, which are straight, than the curved line described by each point of the wheel depends on the straight and circular movements. Accordingly, although it is often useful to divide a movement into several parts, in order to understand it more easily,[23] nonetheless, strictly speaking, we must never attribute more than one movement to each body.

33.   How in all movement a complete circle of bodies moves simultaneously.

It has been shown above [24] that all places are full of bodies and that the size of each part of matter is always exactly equal to that of its place; {so

---

[21] This is true only if one has already chosen a specific frame of reference. If the axle is considered at rest, then a point on the wheel will describe a circle, and the Earth will rotate beneath the wheel in the opposite direction. Of course, in this case the Earth is the most convenient reference frame, for the reasons given in Article 30.

[22] See Plate I, Fig. ii.

[23] The technique of treating a complex motion as being composed of several simpler ones is very ancient. It was the basic technique used for describing planetary motion in both the Ptolemaic and Copernican systems, and was also used by Galileo in discovering the laws of projectile motion.

[24] See Articles 18 and 19.

that it is not possible for it to fill a bigger one or to fit into a smaller one, or for any other body to find room in its place while it is there}. From this it follows that no body can move except in a {complete} circle {of matter or ring of bodies which all move at the same time}; in such a way that it drives another body out of the place which it enters, and that other takes the place of still another, and so on until the last, which enters the place left by the first one at the moment at which the first one leaves it.[25] We easily conceive of this in the case of a perfect circle, because we see that no vacuum and no rarefaction or condensation are required to permit part A[26] of the circle to move toward B, provided that part B moves simultaneously toward C, C toward D, and D toward A. But the same thing can also be understood even in the case of an imperfect circle, indeed, even in an extremely irregular one; provided that we notice the way in which all the inequalities of the spaces can be compensated for by corresponding inequalities in {the} speed {of the parts}. So, without there being any condensation or vacuum; all the matter contained in the space EFGH[27] can move in a circle. The part of it which is near E can move toward G and that which is near G can simultaneously move toward E, provided only that (since we are supposing the space at G to be four times as wide as at E, and twice as wide as at F and H) we also suppose the movement to be four times as rapid at E as at G, and twice as rapid as at F and H. Similarly, in all remaining places, we can suppose that speed of movement compensates for narrowness of space. Thus, in any given length of time, the same quantity of matter will pass through one section of this circle as through another.

34.   That it follows from this that matter is divisible into an indefinite number of parts, even though this is beyond our comprehension.

It must, however, be admitted that there is in this movement something which our mind cannot [fully] understand, even though we perceive it to be true: namely, a division of certain parts of matter to infinity, or an

[25] If all space is completely filled with matter, then a body can move only by pushing adjacent bodies out of its way. Further, since a vacuum is impossible, the space vacated by the body must be filled by other bodies being simultaneously pushed into that space. This means that the motion of the first body must be simultaneously transmitted to those other bodies. Elsewhere, e.g., in the *Dioptrics*, Descartes points out that when one end of a solid rod is moved, the other end moves simultaneously.
[26] See Plate I, Fig. iii.
[27] See Plate I, Fig. iv.

indefinite division into so many particles that we cannot conceive of any so small that we do not understand that it is in fact divided into others even smaller. For it is not possible for the matter which now fills the space G[28] to fill successively all the spaces of very gradually decreasing size which are between G and E, unless some of these parts adapt their shape {and divide as necessary to fit exactly} to the innumerable dimensions of those spaces. In order for this to occur, all the particles into which one can imagine such a unit of matter to be divisible, and which are truly innumerable, must move slightly with respect to one another; and however slight this movement, it is nevertheless a true division.[29]

35.   How this division occurs, and that we must not doubt that it does occur, even though we cannot understand it.

It must be observed that I am not talking here about all matter, but only about some part of it. For although we might suppose that there are, in {the space} G, two or three parts equal in width to space E and an additional number of other smaller ones, which [all] remain undivided; nonetheless we can understand that they may all describe a circular movement in the direction of E, provided that, mingled with those which adapt themselves to the space which must be occupied [in a given time] only by changing the speed at which they travel, there are others which somehow yield and change their shapes in such a way as to exactly fill all the angles {and little corners} which the former group will not occupy.[30] Further, although we cannot comprehend how this indefinite division occurs, we must not on that account doubt that it does occur: because we clearly perceive that it follows necessarily from the nature of matter, which is {already} known to us in a very evident manner; and because we perceive also that this division. is one of those things which cannot be fully grasped by our mind, since our mind is finite.

36.   That God is the primary cause of motion; and that He always maintains an equal quantity of it in the universe.

After having examined the nature of movement, we must consider its cause, which is twofold: {we shall begin with} the universal and primary

---

[28] See Plate I, Fig. iv.

[29] Since there is no vacuum, the only way parts of matter can be divided is for them to change their spatial relationships; this *may* alter the distance between two such parts, but it need not.

[30] Descartes's claim here is simply that while some of the circulating parts of matter may be solid, at least some of the parts must be fluid.

one, which is the general cause of all the movements in the world; and then
{we shall consider} the particular ones, by which individual parts of matter
acquire movements which they did not previously have. As far as the
general {and first} cause is concerned, it seems obvious to me that this is
none other than God Himself, who, {being all-powerful} in the beginning
created matter with both movement and rest; and now maintains in the
sum total of matter, by His normal participation, the same quantity of
motion and rest as He placed in it at that time.[31] For although motion is
only a mode of the matter which is moved, nevertheless there is a fixed and
determined quantity of it; which, as we can easily understand, can be
always the same in the universe as a whole even though there may at times be
more or less motion in certain of its individual parts. That is why we must
think that when one part of matter moves twice as fast as another twice as
large, there is as much motion in the smaller as in the larger; and that
whenever the movement of one part decreases, that of another increases
exactly in proportion. We also understand that it is one of God's
perfections to be not only immutable in His nature, but also immutable and
completely constant in the way He acts. Thus, with the exception of those
changes which either manifest experience or divine revelation renders
certain, and which we either perceive or believe to occur without any
change on the part of the Creator; we must not suppose that there are any
others in His works, for fear of accusing Him of inconstancy. From this it
follows that it is completely consistent with reason for us to think that,
solely because God moved the parts of matter in diverse ways when He first
created them, and still maintains all this matter exactly as it was at its
creation, and subject to the same law as at that time; He also always
maintains in it an equal quantity of motion.[32]

---

[31] It is important to note here that by 'quantity of motion' Descartes does not mean
momentum, i.e., mass times velocity. Rather, he intends quantity of motion to be given by the
product of size (or volume) and speed. This is, of course, a result of his view that extension is the
essential property of matter. Thus, the behavior of bodies should be determined entirely by
their extension, figure, and motion (figure and motion being essential attributes of extended
things). The preference of speed over velocity may result from the claim that the direction in
which a body moves depends upon which other bodies are considered at rest. Therefore, there
is nothing *in the body itself* which enables one to determine its direction of motion.

[32] While this may be consistent with reason, it clearly does not follow. What follows, even on
the most generous interpretation, is that the total quantity of *something* must remain constant.
There was considerable subsequent dispute between the adherents of Descartes and those of
Leibniz as to whether quantity of motion or what we now know as quantity of energy was
conserved. In fact, if quantity of motion is taken to mean momentum, both are conserved.

37.   The first law of nature: that each thing, as far as is in its
            power,[33] always remains in the same state; and that
            consequently, when it is once moved, it always continues to
            move.

Furthermore, from this same immutability of God, we can obtain
knowledge of the rules or laws of nature, which are the secondary and
particular causes of the diverse movements which we notice in individual
bodies. The first of these laws is that each thing, provided that it is simple
and undivided, always remains in the same state as far as is in its power, and
never changes except by external causes. Thus, if some part of matter is
square, we are easily convinced that it will always remain square unless
some external intervention changes its shape. Similarly, if it is at rest, we do
not believe that it will ever begin to move unless driven to do so by some
external cause. Nor, if it is moving, is there any significant reason to think
that it will ever cease to move of its own accord and without some other
thing which impedes it. We must therefore conclude that whatever is
moving always continues to move as far as is in its power. However,
because we inhabit the earth, which is so constituted that all movements
which occur near to it cease in a short while (and frequently from causes
which are concealed from our senses), we often judged, from the beginning
of our life, that those movements which thus ceased for reasons unknown
to us, did so of their own accord. Indeed, because experience seems to have
proved it to us on many occasions, we are still inclined to believe that all
movements cease by virtue of their own nature, or that bodies have a
tendency toward rest. Yet this is assuredly in complete contradiction with
the laws of nature; for rest is the opposite of movement, and nothing moves
by virtue of its own nature toward its opposite or its own destruction.[34]

38.   Why bodies which have been thrown continue to move after
            they leave the hand.[35]

Indeed, daily experience of things which are thrown to a distance

[33] Latin 'quantum in se est'; 'as far as is in its power' or 'as far as it [itself] is concerned'.
[34] Descartes's rejection of this view is associated with his rejection of final causes in general
(see Part I, Article 28). The overthrow of the Ancient Greek view that change was to be
understood in terms of an inner tendency or nature of the changing thing and the realization
that change only results from some sort of interaction were essential ingredients of the
scientific revolution.
[35] This was one of the most difficult problems of Aristotelian physics. Since heavy objects had
an innate tendency toward rest, and since nothing appeared to be pushing a projectile; it was
difficult to explain what was overcoming the body's natural tendency.

confirms this {first} rule in every way. For there is no other reason why
things which have been thrown should continue to move for some time
after they have left the hand which threw them except that, {in accordance
with the laws of nature}, having once begun to move, they continue to do so
until they are slowed down by encounter with other bodies. It is obvious,
moreover, that they are always gradually slowed down, either by the air
itself or by some other fluid bodies through which they are moving, and
that, as a result, their movement cannot last for long. We can in fact prove
by our own sense of touch that the air resists the movement of other bodies,
if we shake an {open} fan vigorously. The flight of birds confirms the same
thing.[36] Moreover, there is no other fluid body {on the earth} which does
not resist the movement of projectiles even more manifestly than does the
air.

39.   The second law of nature: that all movement is, of itself, along
             straight lines;[37] and consequently, bodies which are moving in
             a circle always tend to move away from the center of the circle
             which they are describing.

   The second law of nature {which I observe} is: that each part of matter,
considered individually, tends to continue its movement only along straight
lines, and never along curved ones; even though many of these parts are
frequently forced to move aside because they encounter others in their
path, and even though, as stated before, in any movement, a circle of matter
which moves together is always in some way formed. This rule, like the
preceding one, results from the immutability and simplicity of the
operation by which God maintains movement in matter; for He only
maintains it precisely as it is at the very moment at which He is maintaining
it, and not as it may perhaps have been at some earlier time. Of course, no
movement is accomplished in an instant; yet it is obvious that every moving
body, at any given moment in the course of its movement, is inclined to
continue that movement in some direction in a straight line, and never in a

[36] Presumably, because if the air did not resist the motion of the bird's wings, it would simply
fall to Earth. The French text omits this sentence.
[37] The previous law, combined with the first portion of this one, is generally regarded as the
first statement of what was to become Newton's law of inertia. There is a significant difference
in import between Newton's view and Descartes's, however. Whereas Newton regards motion
and rest as merely quantitatively different; Descartes regards them as *opposite* or opposing
states. See Articles 44, 49, and 50.

curved one. For example, when the stone A is rotated in the sling EA[38] and describes the circle ABF; at the instant at which it is at point A, it is inclined to move along the tangent of the circle toward C. We cannot conceive that it is inclined to any circular movement: for although it will have previously come from L to A along a curved line, none of this circular movement can be understood to remain in it when it is at point A. Moreover, this is confirmed by experience, because if the stone then leaves the sling, it will continue to move, not toward B, but toward C. From this it follows that any body which is moving in a circle constantly tends to move [directly] away from the center of the circle which it is describing. Indeed, our hand can even feel this while we are turning the stone in the sling, {for it pulls and stretches the rope in an attempt to move away from our hand in a straight line}.[39] This consideration {is of such importance, and} will be so frequently used in what follows, that it must be very carefully noticed here; I shall explain it more fully later.

40.   The third law: that a body, upon coming in contact with a stronger one, loses none of its motion; but that, upon coming in contact with a weaker one, it loses as much as it transfers to that weaker body.[40]

This is the third law of nature: when a moving body meets another, if it has less force to continue to move in a straight line than the other has to resist it, it is turned aside in another direction, retaining its quantity of motion and changing only the direction of that motion. If, however, it has more force; it moves the other body with it, and loses as much of its motion as it gives to that other. Thus, we know from experience that when any hard bodies which have been set in motion strike an unyielding body, they do not on that account cease moving, but are driven back in the opposite direction; on the other hand, however, when they strike a yielding body to

[38] See Plate II, Fig. i.

[39] The force described here is known as 'centrifugal force'; a term introduced by Huygens. This force forms the basis of Descartes's planetary mechanics and his explanation of the phenomenon of light in Part III. Unfortunately, with regard to the stone, the force is non-existent. See the extensive commentary to Articles 57 and 58 of Part III.

[40] This apparently innocent law might appear to be a rather trivial consequence of the law of conservation of motion, and it can be so regarded if the meanings of 'stronger' and 'weaker' are left sufficiently vague. When Descartes specifies the conditions under which a body will be weaker or stronger than another, thereby specifying the meaning of those terms, this law becomes one of the principal sources of error in his physics. See esp. Articles 46–52.

which they can easily transfer all their motion, they immediately come to rest. All the individual causes of the changes which occur in [the motion of] bodies are included under this third law, or at least those causes which are physical; for I am not here enquiring into what kind of power the minds of men or Angels may perhaps have to move bodies; I am reserving that matter for a treatise *on man*.

41.   The proof of the first part of this law.

The first part of this law is proved by the fact that there is a difference between motion considered in itself, and its determination in some direction; this difference makes it possible for the determination to be changed while the quantity of motion remains intact. For, as has been stated above, each thing which is not complex but simple, as motion is, always continues to exist as long as it is not destroyed by any external cause. And in an encounter with an unyielding body, there certainly appears a cause which prevents the movement of the body which strikes the other from maintaining its determination in the same direction. However, there is no cause which would remove or decrease the motion itself, {since none is taken from it by this body or any other cause and} since movement is not contrary to movement. From which it follows that its motion must not be diminished.

42.   The proof of the second part.

Similarly, the second part is proved by the immutability of God's manner of working in always uninterruptedly maintaining the world by the same action by which He created it. From the fact that all places are full of bodies and that, nevertheless, the movement of each of these bodies tends in a straight line; it is obvious that when God first created the world, He not only moved its parts in various ways, but also simultaneously caused some of the parts to push others and to transfer their motion to these others. So in now maintaining the world by the same action and with the same laws with which He created it, He conserves motion; not always contained in the same parts of matter, but transferred from some parts to others depending on the ways in which they come in contact. Thus, this continuous changing in created things is an argument for the immutability of God.[41]

---

[41] This is either a straightforward case of affirming the consequent or a vicious circle.

43.   In what the force of each body to drive or to resist consists.

We must however notice carefully at this time in what the force of each body to act against another or to resist the action of that other consists: namely, in the single fact that each thing strives, as far as is in its power, to remain in the same state, in accordance with the first law stated above. From this it follows that a body which is joined to another has some force to resist being separated from it, while a body which is separate has some force to remain separate. One which is at rest has some force to remain at rest, and consequently to resist everything which can change it; while a moving body has some force to continue its motion, i.e., to continue to move at the same speed and in the same direction. Furthermore, this force must be measured not only by the size of the body in which it is, and by the [area of the] surface which separates this body from those around it; but also by the speed and nature of its movement, and by the different ways in which bodies come in contact with one another.

44.   That movement is not contrary to movement, but to rest; and that determination in one direction is the opposite of determination in another.

It must also be noticed that one movement is in no way contrary to another movement of equal speed; but that, strictly speaking, only a twofold opposition is found here. One is between movement and rest, or even between rapidity of movement and slowness of movement (i.e., to the extent that this slowness partakes of the nature of rest): the other is between the determination of a body to move in a given direction and the encounter, in its path, with a body which is either at rest or moving in a contrary manner; and this opposition is greater or smaller according to the direction in which the body which encounters the other is moving.[42]

---

[42] The claim here is that a body has two distinguishable tendencies: a tendency to resist a change from motion to rest or *vice versa*, and a tendency to continue to move in a certain direction. On Descartes's view, the latter tendency is much weaker than the former; which is another serious source of error. See note 49.

45.   How it is possible to determine to what extent the movement of
            each body is changed by coming in contact with other bodies;
            and that this can be done according to the following rules.[43]

In order to determine, from the preceding laws, how individual bodies
increase or decrease their movements or turn aside in different directions
because of encounters with other bodies; it is only necessary to calculate
how much force to move or to resist movement there is in each body; and to
accept as a certainty that the one which is the stronger will always produce
its effect. Moreover, this could easily be calculated if only two bodies were
to come in contact, and if they were perfectly solid,[44] and separated from
all others {both solid and fluid} in such a way that their movements would
be neither impeded nor aided by any other surrounding bodies;[45] for then
they would observe the following rules.

46.   The first rule.[46]

First, if these two bodies, for example B and C,[47] were completely equal
in size and were moving at equal speeds, B from right to left, and C toward
B in a straight line from left to right; when they collided, they would spring
back and subsequently continue to move, B toward the right and C toward

---

[43]The following rules all presuppose conservation of motion, the general equation being:
Bb + Cc = Bb′ + Cc′; where B and C are the volumes of the respective bodies, b and c their
initial speeds, and b' and c' the resultant speeds. Clearly, for any initial set of volumes and
speeds, the equation will have an infinite number of solutions. The third law, Article 40, is
designed to overcome this difficulty. That law was disproved by Huygens, and the correct laws
of elastic impact were submitted by him to the Royal Society in 1669.

[44] Presumably, 'elastic' is meant here; the Latin term is *durus*, which means 'hard', 'solid', or
'unyielding'.

[45].The intended procedure here is analogous to that of Galileo in determining the law of falling
bodies; that is, to determine how bodies would behave under ideal conditions and then to take
account of disturbing factors such as air resistance, shape, etc. Of course, Descartes considers
a vacuum logically impossible, which raises serious questions about the status of his rules.

[46] There is evidence to suggest that Descartes himself is responsible for the additional
arguments and examples in the French version of Articles 46–52. The evidence consists of
Burman's account of a conversation he had with Descartes on April 16, 1648. Describing
Descartes's reply to some question Burman had raised about the first rule, Burman quotes:
"The author has, in the French *Principles*, to some extent elucidated and explained these laws,
because many have been complaining of their obscurity." See A. & T., V, 168; also, see note
62.

[47] Articles 46 through 52 refer to Plate II, Fig. ii.

the left, without having lost any of their speed.[48] {For, in this case, there is no cause which could take their speed from them, but there is a very obvious one which must force them to spring back; and because it would be equal in each, they would both spring back in the same way}.

47.   The second.

Second, if B were slightly larger than C, and everything else were as previously described, then only C would spring back, and both would move toward the left at the same speed.[49] {For B, having more force than C, could not be obliged by C to spring back}.

48.   The third.

Third, if the two bodies were equal in size, but if B were moving slightly more rapidly than C; after their collision not only would {C alone spring back and} both continue their movement toward the left, {that is, in the direction from which C came}, but also one half of B's additional speed would be transferred from it to C, {since B could not move more rapidly than C which would be ahead of it}. For example, if B had initially been travelling at six degrees of speed [toward the left], and C at a speed of only four [toward the right], {B would transfer to C one of its two additional degrees of speed, and} both would subsequently move toward the left at five degrees of speed.[50] {This would occur because it is much easier for B to transfer one of its additional degrees of speed to C than for C to change the course of all the movement which is in B}.[51]

[48] If one assumes equal masses, this is the correct result. Indeed, whenever two equal elastic masses collide in a straight line, they simply *exchange* speeds and directions; i.e., velocities.
[49] Assuming that size and mass are proportional, this result is incorrect. What actually results depends entirely upon the ratio of the two masses. In general, however, B will lose speed, perhaps springing back or coming to rest; and C will change direction and gain speed. The result Descartes describes never occurs, regardless of the ratio of B's mass to that of C. For example, if B has twice the mass of C, and each is moving with a speed of 1.5, then in fact both will change direction. B will have a subsequent speed of 0.5 and C a speed of 2.5 in the opposite direction. Further, it is apparent that the second rule implies that the subsequent speed of both B and C must be the same as their initial speed, otherwise motion would be either gained or lost. This is a striking illustration of Descartes's view that change in direction is much easier to produce than change in motion, since B reverses C's direction of motion without being itself affected in any way.
[50] This too is incorrect; see note 48.
[51] This is presumably intended to follow from the third law; but since the law provides no quantitative definition of force, the conclusion here is gratuitous.

49.   The fourth.[52]

Fourth, if the body C were entirely at rest, {that is, if it not only had no
apparent motion but also were not surrounded by air or any other fluid
(which makes the hard bodies immersed in such a fluid very easily movable,
as I shall show)}, and if C were slightly larger than B; the latter could never
{have the force to} move C, no matter how great the speed at which B might
approach C. Rather, B would be driven back by C in the opposite
direction: because {for B to move C, C would have to be driven as rapidly as
B subsequently moves and} a body which is at rest puts up more resistance
to high speed than to low speed; and this resistance increases in proportion
to the difference in the speeds.[53] Consequently, there would always be more
force in C to resist than in B to drive, {because C is larger. For example, if B
is one half as large as C and is travelling at three degrees of speed, then
because B is only as large as each of C's halves and because it cannot
continue in the same direction more rapidly than it pushes C ahead of it; B
cannot move C without transferring to it two thirds of its quantity of
motion, one third for each of C's halves, keeping for itself only one degree
of speed.[54] Similarly, if B has thirty degrees of speed, twenty will have to be
communicated to C; if B has three hundred, two hundred will have to be
transferred, and so on. But since C is at rest, its resistance to receiving
twenty degrees of speed is ten times as great as its resistance to receiving
two, and so on. Thus, the greater B's speed, the proportionally greater C's
resistance will be. And because each half of C has as much force to remain
at rest as B has to drive it, and because both halves resist at the same time, it
is obvious that they must succeed in forcing B to spring back. So that, no
matter how great the speed at which B approaches C, B can never have the
force to move C}.[55]

[52]This rule and the following one are most interesting. They illustrate Descartes's view that
motion and rest are opposing or opposite states, and his view that resistance to motion
depends entirely on relative size. Quantity of motion plays no role whatever, except that it
must be conserved.

[53] Even if this were so, it clearly does not follow that the resistance will be *greater* than the
force of B.

[54] From the assumption that B can move C only if they subsequently move in the same
direction at the same speed and from conservation of motion, it follows that they must both
move at a speed of one, C having twice B's quantity of motion; making the total quantity of
motion equal three. In fact, B recoils with a speed of one, and C moves off with a speed of two.
This preserves quantity of momentum, since B's velocity is now − 1, but gives a quantity of
motion of five on Descartes's view.

[55] The additional material from the French text has been extensively rewritten in an attempt to
make it more clear.

50.   The fifth.

Fifth, if the body C were at rest and {even very slightly} smaller than B;
then, no matter how slowly B might advance toward C, it would move C
with it by transferring to C as much of its motion as would permit the two to
travel subsequently at the same speed.[56] Thus if B were twice as large as C, it
would transfer to C {only} one third of its quantity of motion; because that
one third would move the body C at the same speed as the remaining two
thirds would move the body B which {we are supposing} is twice as large as
C. Therefore, after B had collided with C, its speed would be reduced by one
third; that is to say, B would then need as much time to travel a distance of
two feet as it previously did to travel a distance of three feet.[57] Similarly, if
B were three times as large as C, it would transfer to C one quarter of its
motion; and so on. {And it is impossible for B to have so little force that it
would ever be insufficient to move C; for it is certain that weaker motions
must observe the same laws as stronger ones, and must produce,
proportionally, the same [type of] result. Although we often think we see
the opposite on this earth; this is because of the air and other fluids which
always surround solid moving bodies and which can greatly increase or
decrease their speed, as we shall see later}.

51.   The sixth.

Sixth, if the body C were at rest and exactly equal in size to body B,
which was moving toward it; necessarily, C would be to some extent driven
forward by B and would to some extent drive B back in the opposite
direction. Thus, if B were to approach C with four degrees of speed, it
would {have to} communicate one degree to C, and be driven back in the

---

[56] This rule seems to be simply the converse of the preceding. The pair, however, are in serious
conflict with Descartes's principle of the relativity of motion. Since by hypothesis there are no
"immediately contiguous bodies" except B and C involved; the antecedent conditions
described in the two rules are the same; simply depending on whether B or C is considered to
be at rest. (Indeed, the conditions described in rule two should also constitute a correct
description of the same situation.) Thus, the results should be the same in each case.
[57] Assuming that B has three degrees of speed, as in the example in the previous rule; the
correct result is that B will retain its direction with a speed of one, and C will acquire a speed of
four. Notice that in both these cases, and in the example given in note 49, the *relative* speed
between B and C remains constant. Before collision they move toward one another with a total
speed of three; after collision, they move away from one another with the same total speed.

opposite direction with the remaining three.[58] {Because, it must necessarily be the case that either B moves C without springing back, thus transferring two degrees of its speed to C; or that B springs back without moving C, retaining those two degrees of speed as well as the two which cannot be taken from it; or else that B springs back, retaining some portion of those two [extra] degrees of speed, and at the same time moves C with the remainder of those two degrees. Since B and C are equal and there is consequently no more reason for B to spring back than to move C; it is obvious that these two effects must be equally shared: that is, B must transfer one of these degrees of speed and spring back while retaining the other}.

52.   The seventh.

Finally, if B and C were travelling in the same direction, C more slowly than B, so that B (which would be following C) would eventually strike it; and if C were larger than B but B's speed exceeded C's by a greater extent than C's size exceeded B's: then B would transfer to C as much of its speed as would be required to permit them both to travel subsequently at the same speed and in the same direction. However, if, on the contrary, B's speed exceeded C's by a smaller extent than C's size exceeded B's; B would be driven back in the opposite direction, and would retain all its movement.[59] {And, finally, when the ratio in which C's size exceeds B's is exactly equal to the ratio in which B's speed exceeds that of C, B must transfer some of its motion to C and spring back with the rest}.[60] The effect of the extent to which these ratios exceed each other is calculated as follows: if C were twice as large as B, and if B were not moving twice as rapidly as C, B would not drive C forward but would be driven back in the opposite direction; if, however, B were moving more than twice as fast as C, it would drive C forward with as much of its motion as is required to cause both to move at the same speed. Thus, if C had only two degrees of speed, and B had five: two degrees of speed would have to be taken away from B,

---

[58] This rule seems to result from combining the two previous rules. Since C is neither larger nor smaller than B; there will be some tendency for C to react as if it were smaller than B, and for both to move with a speed of two (rule five). However, there will be an equal tendency for C to behave as if it were larger and hence for it to acquire no speed whatever (rule four). Taking the average of these two tendencies gives the result. For the correct result, see note 48.

[59] These two rules seem to be generalizations of rules five and four, respectively.

[60] As with rule six, this added rule seems to represent a combination of the two previous situations.

and once transferred to C would form only one degree, since C is twice as large as B: as a result, the two bodies B and C would each subsequently travel with three degrees of speed;[61] and so on. These things require no proof, because they are obvious in themselves.[62]

53.   That the application of these rules is difficult, because each body is always surrounded by many contiguous ones.

{Indeed, experience often seems to contradict the rules I have just explained}. However, because there cannot be any bodies in the world which are thus separated from all others, and because we seldom encounter bodies which are perfectly solid; it is very difficult to perform the calculation to determine to what extent the movement of each body may be changed by collision with others. Since, {before we can judge whether these rules are observed here or not}, we must simultaneously calculate the effects of all those bodies which surround the bodies in question and which affect their motion. These effects differ greatly, depending on whether the surrounding bodies are solid or fluid; and it is therefore necessary that we should immediately enquire into the difference between solid and fluid bodies.

54.   What solid and fluid bodies are.

Of course, from the testimony of our senses, {for these properties are in their domain}, we recognize this difference to consist simply in the fact that the parts of fluid bodies easily move out of their places, and consequently do not resist the movement of our hands into those places; while, on the contrary, the parts of solid bodies adhere to one another in such a way that,

[61] In fact, C will increase its speed to four and B will be slowed to a speed of one; retaining the relative speed of three between B and C.

[62] In the French text, the final sentence is: "And the demonstrations of this are so certain that, even if experience were to appear to show us the opposite, we would nevertheless be obliged to place more trust in our reason than in our senses." The editors of the re-edition of Adam and Tannery assert that Descartes was responsible for this alteration as well as for the expanded versions of the rules of collision in the French text. The evidence for this is a letter from Descartes to Mersenne dated April 20, 1646. Descartes writes, "If you see M. Picot [the author of the French translation], please tell him that I have received his letters but that I cannot yet send him the continuation of his translation, because I have not yet succeeded, in the entire year which has passed since I reached that article, in finding a few moments in which to clarify my laws of movement.": A. & T., IV, 396.

without sufficient [external] force to overcome their cohesion, they cannot be separated. And upon further investigation into how it happens that some bodies give up their places to others without any difficulty, while other bodies do not do so, we easily notice that those which are already in motion do not prevent the places which they are leaving of their own accord from being occupied by others; but that those which are at rest cannot be driven out of their places without some {external} force {causing this change}. From this we may conclude that those bodies which are divided into very small parts which are agitated by a diversity of {independent} movements, are fluid; while those bodies whose particles are all contiguous and at rest, are solid.

55.   That the parts of solid bodies are not joined by any other bond than their own rest {relative to each other}.

Furthermore, our reason certainly cannot discover any bond which could join the particles of solid bodies more firmly together than does their own rest. For what could this bond be? It could not be a substance, because there is no reason why these particles, which are substances, should be joined by any substance other than themselves.[63] Nor is it a mode[64] different from rest; for no other mode can be more opposed to the movement which would separate these particles than is their own rest. Yet, besides substances and their modes, we know no other kinds of things.[65]

56.   That the particles of fluids {tend to} move with equal force in all directions; and that a solid.body, immersed in a fluid, can be set in motion with very little force.

However, as far as fluid bodies are concerned, even though our senses may not inform us that their particles move, since they are too small, this is nonetheless easily deduced from effects; especially in the cases of air and water, because many other bodies are destroyed by them: for no physical

[63] Presumably because the solidity of the bonding substance would itself need to be explained.
[64] The French has 'quality' here and in the remainder of the article.
[65] This is not, of course, an explanation of solidity but a *description* of the fact that the parts of a solid do not move relative to one another. Newton, in Query XXXI of his *Opticks*, states: "And for explaining how this [solidity] may be, some have invented hooked Atoms, which is begging the Question; and others tell us that Bodies are glued together by rest, that is by an occult Quality, or rather by nothing; and others, that they stick together by conspiring Motions [of their parts], that is, by relative rest amongst themselves."

action of that kind can occur without {the fluid particles} moving [relative to each other].[66] The causes of these movements will be indicated later.[67] Yet there is a difficulty here, because these particles of fluids cannot all move at the same time in every direction; which appears to be required if they are not to impede the movement of bodies coming from any direction. Indeed, we see that they do not impede the movement of bodies. For example, if the solid body B is moving toward C[68] while some parts of the fluid D, which are between B and C, are moving in the opposite direction from C to B; these will not aid B's movement, but will on the contrary impede it more than if they were completely at rest. In order to resolve this difficulty, we must recollect that it is not movement but rest which is contrary to movement; and that the determination of a movement in one direction is contrary to its determination in the opposite direction, as was stated earlier. Furthermore, [we must remember that] all moving bodies always tend to continue their movement in a straight line. Now from these things, it is obvious: first, that when the solid body B is at rest, it puts up more opposition, by its rest, to the movements of the particles of the fluid body D considered collectively, than it would by its movement if it were moving. And second, that as far as determination is concerned, the fact is that there are as many particles of D moving from C to B as in the opposite direction: for the same ones which come from C strike the surface of the body B and are driven back toward C. Although some of these particles, considered individually, strike B and drive it toward F (and thus more greatly impede its movement toward C than if they were at rest); an equal quantity of particles also move from F toward B, and drive B toward C. The result is that B is no more driven in one direction than in another, and therefore remains at rest, unless something else intervenes. No matter what we suppose the shape of B to be, it is always driven by exactly the same number of particles of the fluid coming from one direction as from the other;[69] provided that the fluid itself is not moving in any one direction more than in the others {like that of a river}. We must also suppose B to be surrounded on all sides by the fluid DF; and if it happens that there is not as great a quantity of this fluid at F as at D, this does not matter: because the

---

[66] The Latin simply says, "...without local motion...," here.

[67] See Part III, Articles 49–51.

[68] Articles 56 through 60 refer to Plate II, Fig. iii.

[69] Presumably, the intention here is to claim that the force of the particles which incline the body in one direction is equal to the force inclining it in an opposite direction. The number of particles would not be equal unless the surface area of the body were equal in all directions.

fluid does not act as a whole against B, but only [acts] with those of its parts which touch B's surface. Thus far, however, we have been considering B to be at rest; if we now suppose it to be driven toward C by some force coming from elsewhere, this force, however slight, would suffice, not indeed to move B by itself, but to unite with the [force of the] particles of the fluid body FD and to enable them to drive B toward C and to transfer to B some of their motion.

57.   The proof of this matter.

In order that this may be more clearly understood, let us first suppose that the solid body B is not yet in the fluid FD, but that the particles *aeioa* of this fluid, arranged in the form of a ring, are moving circularly in the order of the symbols *aei*; and that others *ouyao* are moving similarly in the order of the symbols *ouy*. For in order for any body to be fluid, its particles must move in many {diverse} ways, as has just been stated. Then, if the solid body B is at rest in this fluid FD, between *a* and *o*, what will happen? The particles *aeio* will certainly be prevented by B from moving from *o* toward *a* to complete the circle of their movement; and similarly, the particles *ouya* will be prevented from continuing from *a* toward *o*. Those coming from *i* toward *o* will drive B toward C, while those coming from *y* toward *a* will drive it back equally toward F. As a result, these particles alone will have no force to move B, but will be driven back from *o* toward *u*, and from *a* toward *e*; and one circulation will be formed from two, following the order of the symbols *aeiouya*. Thus, collision with the body B will not in any way affect the [quantity of] motion of these particles, but will only change their determination; so that they will not move along such straight lines, or along lines so close to straight, as if they had not struck B. Then, finally, if some external force intervenes, driving B toward C; this force, however slight, joined to that by which the particles of the fluid coming from *i* toward *o* also drive B toward C, will overcome that by which the particles coming from *y* toward *a* drive B back in the opposite direction; and will therefore suffice to change their determination and to cause them to travel in the order of the symbols *ayuo* to the extent required for the movement of body B not to be impeded: because, when two bodies are determined to move in completely opposite directions, the body which has the greater force must change the determination of the other. Moreover, what I am saying here about the particles *aeiouy* is to be understood also of all the other particles of the fluid FD which strike B: those particles which are driving B toward C are opposed by an equal number of others driving B in the opposite direction; and even a very slight force, united with that of some of these particles, will suffice to change the determination of the opposing particles. And although these particles may perhaps not be describing circles like those illustrated here, there is nevertheless no doubt that they all move circularly in other ways equivalent [to those shown].[70]

58.   That if these particles of the fluid are moving more slowly than the solid body situated in it, that portion of it does not behave like a fluid.

Now, when the determination of the particles of the fluid which were preventing the body B from moving toward C has thus been changed, the body B will begin to move at the same speed at which it is driven by the external force, provided that all of the particles in this fluid are moving at least as rapidly as the speed of that force.[71] For if some particles are moving more slowly, to the extent that it is composed of them, the fluid does not behave entirely like a fluid; nor will the slightest force then be sufficient to move a solid body immersed in such a fluid. Rather, a force great enough to overcome the resistance arising from the [comparative] slowness of the particles of the fluid is now required. Thus we often see that air, water, and other such fluids, put up much resistance to bodies which are moving very rapidly through them, yet yield without any difficulty to these same bodies when they are moving more slowly.

59.   That a solid body which has been driven by another does not receive all its movement from that other, but also acquires some motion from the surrounding fluid.

However, when the body B is thus moving toward C, it must not be thought that B acquires its motion solely from the external force driving it. It {also} acquires motion to a great extent from the particles of the fluid; so that the particles forming the circles *aeio* and *ayuo* lose as much of their movement as is acquired by those particles of the solid body B which are between *o* and *a*;[72] since these particles will now form a part of the circular

---

[70] That is, regardless of which *particular* circulations happen to be present in a fluid, the net result will be as described, and the immersed body will, if left to itself, always be in equilibrium.
[71] "... at least as rapidly as the external force is moving B." seems to be meant here.
[72] Presumably, the parts of the fluid ahead of B will gain motion by being pushed by B; although Descartes seems unclear on this point.

movements *aeioa* and *ayuoa*: although, as they advance further toward C, they are constantly united with new particles of the fluid.[73].

60.   That a solid body cannot, however, acquire a greater speed
                from this fluid than it has acquired from the solid body which
                drives it.

It only remains for me to explain here why I did not state a little earlier that the determination of the particles *ayuo* was completely changed, but only that it was changed to the extent required for the movement of the body B to be unimpeded. For in fact, this body B cannot move more rapidly than it has been driven by the external force; although all the particles of the fluid FD may often have much more agitation. This is one of the things which we must especially observe while we are philosophizing: we must never attribute to a cause any effect which exceeds its capacity. Thus, suppose that the body B, formerly immobile and immersed in the fluid FD, is now slowly driven by some external force, for example by my hand. Since the impulse of my hand is the sole cause of its movement[74], we must not believe that B moves more rapidly than it is driven. And although all the particles of the fluid may be moving much more rapidly [than B], it must not be thought on that account that they are determined to circular movements like *aeioa* and *ayuoa*, at a speed greater than that of the force {driving B}; but rather, that insofar as they are more rapidly agitated, they {employ their additional agitation to} move in all other directions, as formerly.[75]

---

[73] I.e., particles which have been displaced by B.

[74] The hand is not the sole cause, of course, but it is what makes the difference between rest and motion; cf. Article 59.

[75] The apparent confusion in Descartes's entire theory of fluid motion is in part unavoidable. He wishes to claim that a body immersed in a fluid offers no resistance to motion, since this is one of the necessary foundations of his vortex theory of planetary motion. This seems, however, to contradict his own rules of impact, especially rule four. Thus, he holds that a body immersed in a highly agitated fluid is subjected to forces from all sides, each of which would be sufficient to move the body were it not for the equal opposing force of the fluid on the opposite side of the body. The opposing force is easily overcome, however, since it involves only a change in the *direction* of motion of the opposing particles. For example, in a letter written to de Beaune, in April, 1639, Descartes says: ". . . thus, when a stone falls to earth from a high place, if it stops and does not roll, I conceive that this is because it shakes the earth and thus transfers its motion to it. But if the quantity of earth it moves contains a thousand times as much matter as it does, when it transfers all its motion to the earth, it gives to it only a thousandth part of its speed." Curiously, an identical passage occurs in a letter whose date is given as 1648. See A. & T., II, 543; and V. 133–139.

61.   That when an entire fluid body moves simultaneously in some
        direction, it must necessarily carry along with it any solid body
        which is immersed in it.

From the preceding, it is clearly perceived that a solid body, immersed in
a fluid and at rest in it, is held there as if in equilibrium. Further, no matter
how large it may be, it can always be driven in one direction or another by
the least force; whether this force comes from elsewhere, or whether it
consists in the fact that this entire fluid simultaneously moves in a certain
direction; as rivers flow to the ocean, or as all the air flows toward the West
when the East wind blows. When this occurs, it is absolutely necessary for a
solid body situated in such a fluid to be carried along with it: nor is this
contradicted by the fourth rule; according to which, as I stated before, a
body which is at rest cannot be set in motion by any smaller than itself, no
matter how rapidly the smaller body may be moving.

62.   That a solid body, which is thus carried along by a fluid, is not
        therefore moving.[76]

If, moreover, we turn our attention to the true and absolute nature of
movement; which consists in the transfer of a moving body from the
vicinity of the other bodies contiguous to it, and which is equal in both the
body which is said to move and the contiguous body away from which [it is
said that] it moves, although it is not customary to speak of the two in the
same way {and to say that both move}: we will clearly know that a solid
body which is thus carried along by the fluid in which it is contained does
not, strictly speaking, move as much as it would if it were not carried along
by this fluid; for it certainly moves away less from the neighboring
particles of this fluid {when it follows its current than when it resists it}.

63.   Why some bodies are so solid that, although small, they cannot
        easily be divided by our hands.

There remains one thing in which experience seems to strongly
contradict the rules of movement which were propounded a short while
ago; for we see many bodies much smaller than our hands, [the particles of]

[76] This is a crucial claim for Descartes. It is on the basis of this principle that he feels he can
adopt the Copernican system in Part III, and still maintain that the Earth does not move;
thereby avoiding any suspicion of heresy.

which adhere together so firmly that they cannot be divided by any force of
our hands. Now, if their parts are joined by no other bond than the fact that
they are contiguous and at rest, and since any body which is at rest can be
set in motion by a moving body which is larger than itself;[77] at first glance
there seems to be no reason why, for example, an iron nail (or any other
body which is not large but extremely solid), cannot be divided into two
parts solely by the force of our hands. For each half of this nail may be
considered to be an individual body; and since one half is smaller than our
hand, it seems that it ought to be possible to move it by the force of our
hand and thus separate it from the other half. It must, however, be noted
that our hands are extremely yielding, or closer to the nature of fluid bodies
than to that of solid ones; for that reason, they are not accustomed to act as
a whole against a body which they have to move; only that part of our
hands which touches that body brings all its pressure to bear upon it at the
same time. For in fact, just as one half of the iron nail (inasmuch as it is to be
divided from the other half), has the nature of an individual body; so also
the part of our hand which is immediately touching it, and which is smaller
than it is, has the nature of an individual body (inasmuch as it can be
separated from the other remaining parts of this hand). It is because this
part can more easily be separated from the rest of the hand than the part of
the nail from the rest of the nail, and because this separation cannot occur
without the sensation of pain, that we cannot break the iron nail by means
of our hand alone. If, however, in order to divide the body, we strengthen
our hand by applying the force of a hammer, file, pair of cutters, or other
tool to a part of the body to be divided which is smaller than the tool being
used; it will be easy to overcome its hardness.

64.   That I do not accept or desire in Physics any other principles
             than in Geometry or abstract Mathematics; because all the
             phenomena of nature are explained thereby, and certain
             demonstrations concerning them can be given.

I shall not add anything here concerning figures, or the way in which
there also result, from their infinite diversity, innumerable diversities of
movement; because these things will be, of themselves, sufficiently obvious
when the occasion to discuss them arises. Furthermore, I am supposing
that my readers are already familiar with the rudiments of Geometry, or

[77] Rule five.

that they at least have capacities adequate to the understanding of Mathematical demonstrations. For I openly acknowledge that I know of no kind of material substance other than that which can be divided, shaped, and moved in every possible way, and which Geometers call quantity and take as the object of their demonstrations. And [I also acknowledge] that there is absolutely nothing to investigate about this substance except those divisions, shapes, and movements; and that nothing concerning these can be accepted as true unless it is deduced from common notions, whose truth we cannot doubt, with such certainty that it must be considered as a Mathematical demonstration. And because all Natural Phenomena can thus be explained, as will appear in what follows; I think that no other principles of Physics should be accepted, or even desired.

PART III

OF THE VISIBLE UNIVERSE

PART III

TRANSLATORS' INTRODUCTION

Part III of the *Principles* represents Descartes's views on astronomy, and might be better entitled "On the Heavens". Since the *Principles* was written during a period of intense intellectual turmoil and change, a process now known as the Copernican Revolution; some knowledge of both the state of astronomy and the general intellectual climate of Descartes's time is required for an understanding of the issues being dealt with in Part III.

The astronomy of the late-sixteenth and early-seventeenth centuries is dominated by four figures. In 1543, Nicolaus Copernicus (1473–1543) published *De Revolutionibus Orbium Coelestium*, setting forth a new proposal for the arrangement of the universe. Previously, the accepted astronomical theory had been that of Claudius Ptolemy (ca. 127–150 A.D.), the last great Ancient astronomer, in which the stars, planets, moon, and sun all revolved around a central stationary Earth: see Plate III. The system was extremely complex, and required a number of interdependent circular motions to account for the observed movements of the heavenly bodies. The Ptolemaic system, however, reflected the major features of Aristotle's physics, metaphysics, and cosmology; thus, to reject it required the abandonment or extensive revision of the entire Aristotelian world picture. Copernicus suggested that the Earth had a daily rotation around its axis, thereby accounting for the apparent daily revolution of the heavens; and an annual revolution around the sun, thereby accounting for the sun's apparent annual journey around the zodiac: see Plate IV. Although Copernicus' own system was extremely complex, it had two great advantages. First, by having the planets also revolve around the sun, Copernicus was able to show that the gross irregularities in the apparent motions of the planets, such as the fact that the planets sometimes appear to reverse their course and move backwards, were *only* apparent and were due to the fact that the planets were being observed from a moving Earth. Second, a great many apparently accidental features of planetary motion, such as the fact that Mercury and Venus always appear very close to the sun, were now explained as inevitable consequences of the new model.

The second figure is Tycho Brahe (1546–1601), the greatest astronomical observer of his time, perhaps the greatest of any time. Although an opponent of the Copernican system, he made three important contributions to its eventual acceptance. First, by observing the nova, or "new star", of 1572 and determining that it was located well beyond the orbit of the moon, he refuted the Aristotelian view that the heavens were perfect and unchanging. Second, he determined that the comet of 1577 was located in the heavens, and thus that comets were heavenly bodies rather than terrestrial, atmospheric phenomena. Since the comet moved *through* the heavens, with a path and motion unique to it, this refuted the main feature of Aristotelian cosmology in which the heavenly bodies were carried around by a series of hollow, spherical, concentric shells of impenetrable crystalline aether in which they were embedded. Finally, Tycho provided subsequent astronomers with a huge body of observational data which were accurate almost to the limits of naked-eye observation. This freed astronomers from their dependence on Ancient observations, many of which were simply incorrect; and provided Kepler with the material on which his contributions were based.

Johann Kepler (1571–1630) modified Copernicus' complex system into, essentially, its final form. This was largely achieved by his discovery, in 1609, of the first two laws of planetary motion. First, that the planets, including the Earth, revolve around the sun in elliptical orbits with the sun at one focus of the ellipse. Second, that the speed of each planet varies with its distance from the sun in such a way that a line joining the planet and the sun will sweep over equal areas of the planet's orbital plane in equal times. The third law, stating that the ratio of the cube of a planet's average distance from the sun to the square of its orbital period is the same for all planets, was discovered in 1619. With these laws, it was possible, for the first time, to make very accurate predictions of planetary positions. Kepler did just this in 1627 when he issued the *Rudolphine Tables*, which were vastly superior to any previous astronomical tables. Thus, after 1627, Kepler's version of Copernicus' system was used by virtually all astronomers, whether or not they professed belief in its physical reality.

Finally, Galileo Galilei (1564–1642) made the first telescopic observations of the heavens in 1609. Using a telescope of his own construction, he made a host of discoveries, including sunspots, mountains and valleys on the moon, countless new stars, the phases of Venus, and the four largest satellites of Jupiter. His discoveries were announced in *Nuncius*

Sidereus, published in 1610. The work was extremely popular, and turned astronomical theory from a highly restricted speciality into a matter of general speculation. More importantly, it showed that the choice of an astronomical theory was not solely dependent on metaphysics or theology, but one on which direct empirical evidence could be brought to bear. Virtually all of Galileo's discoveries refuted or brought into question some feature of the traditional view.

However, in 1616, the Congregation of the Holy Office declared *De Revolutionibus* ... "contrary to scripture", and issued a decree in 1620 forbidding "... all other books teaching the same thing." In 1632, Galileo published his *Dialogue Concerning the Two Chief World Systems*, which purported to be an impartial examination of the merits of the Ptolemaic and Copernican views, but which in fact was thoroughly pro-Copernican. The work was placed on the *Index* of proscribed works in 1633, and Galileo was condemned and forced to recant his views. Descartes was very distressed by this; since at that time he was preparing a work entitled *The World, or A Treatise on Light*, which advocated the Earth's motion. Part III of the *Principles* represents Descartes's attempt to reconcile his own astronomical views with the views of the Church. His letter to Mersenne of November, 1633, reveals Descartes's deep concern and his desire to avoid any appearance of denying Church Doctrine.

... In fact I had decided to send you my *World* as a New Year's gift; and no more than a fortnight ago I was still determined to send you at least a part of it, if the whole could not be transcribed in that space of time. But I shall tell you that, after I had someone enquire in Amsterdam and Leiden ... whether Galileo's *World System* was available; ... I was informed that it had indeed been printed, but that all the copies had been simultaneously burned in Rome and Galileo himself subjected to some sanction. This astonished me so much that I have more or less decided to burn all my papers, or at least to permit no one to see them. For I could not imagine that an Italian, especially one who is, so I hear, well-considered by the Pope; could have been condemned for anything other than the fact that he doubtless attempted to establish the Earth's motion. I well know that this view was formerly censured by some Cardinals, but I thought I had heard that it was being taught publicly, even in Rome; and I confess that if it is false, so are all the foundations of my Philosophy, since they clearly demonstrate this motion. And it is so connected to all the [other] parts of my *Treatise*, that I cannot omit it without rendering the remainder completely defective. But since I would not wish, for anything in the world, to write a discourse containing the slightest word which the Church might disapprove; I would, therefore, prefer to suppress it, rather than publish it in a mutilated version. (A. & T., I, 270–272)

PART III

1. 　　　　That the works of God cannot be thought too great.

We have discovered certain principles concerning material things; and there can be no doubt about the truth of these principles, since we sought them by the light of reason and not through the prejudices of the senses. We must now consider whether we are able to explain all the phenomena of nature by these principles alone; and we must begin with those phenomena which are the most universal and on which the rest depend, namely, the general structure of this whole visible world. In order to reason correctly about this matter, we must pay special attention to two things. First, remembering God's infinite power and goodness, we must not be afraid of overestimating the greatness, beauty, and perfection of His works; rather, we must beware of accidentally attributing to them any limits of which we do not have certain knowledge, and of thus seeming to have an inadequate awareness of the Creator's power.[1]

2. 　　　　That we must beware, lest, thinking too highly of ourselves, we suppose that we understand for what ends God created the world.

Second, we must beware of overestimating ourselves. We would be doing so if we were to attribute to the universe limits of which we had not been assured either by reason or by divine revelation; for this would be to assume that our minds can conceive something which is greater than the world which God actually created. We would be overestimating ourselves still more if we were to imagine that God created all things solely for us, or if we were to consider our intellect powerful enough to understand His ends in creating the universe.

---

[1] Belief in an infinite universe, with the stars being vastly remote suns, became an almost integral part of the Copernican view; although Copernicus himself rejected it. The *Principles* seems to have played an important role in this development. As T.S. Kuhn notes in *The Copernican Revolution* (Cambridge, Harvard, 1957), p. 289: "From Bruno's death in 1600 to the publication of Descartes's *Principles of Philosophy* in 1644, no Copernican of any prominence appears to have espoused the infinite universe, at least in public. After Descartes, however, no Copernican seems to have opposed the conception."

3.   In what sense it may be said that all things were created for man.

For although, from a moral point of view, it may be a {good and} pious thought to believe that God created all things for us, since this may move us all the more to love Him and to give thanks to Him for so many blessings; and it is true in a sense, since there is no created thing which we cannot put to some use, even if it is only a matter of exercising our minds by contemplating it and, by means of it, being moved to praise God: it is, however, in no way likely that all things were made for us in the sense that God had no other purpose in creating them. And {it seems to me that} it would be clearly ridiculous to attempt to use such an opinion to support reasonings about Physics; for we cannot doubt that there are many things which are currently in the world, or which were formerly here and have already entirely ceased to exist, which no man has ever seen or known or used.

4.   Of phenomena or experiments and of their use in philosophy.

However, the principles which we have already discovered are so vast and fertile that many more things follow from them than we see included in this visible universe, and even many more than we could mentally examine {in our entire lives}. But let us now set forth a brief description of the principal natural phenomena whose causes are to be investigated here, though not in order that we may use them to prove anything. For we wish to deduce the effects from their causes rather than the causes from their effects. Rather, [we do this] only so that we can consider some, rather than others, of the innumerable effects which we judge can be produced by those causes.

5.   The ratios of distance and magnitude between the Sun, the
            Earth, and the Moon.

Our initial impression is that the Earth is much larger than all the other bodies in the world, and that the Moon and Sun are much larger than the other stars. But if we correct the mistaken impression of our sight by some infallible reasoning, we shall conclude that the Moon is separated from us by a distance of about thirty times the Earth's diameter, and the Sun, by a distance six or seven hundred times that diameter. And if we compare their distances with the apparent diameters of the Sun and Moon, we shall conclude that the Moon is much smaller than the Earth, and the Sun much larger.

6.   The distance between the other Planets and the Sun.

We shall learn also, by means of observation aided by our reason, that the distance from Mercury to the Sun is more than two hundred times the Earth's diameter; from Venus, the distance is more than four hundred times that diameter; from Mars, nine hundred or a thousand times; from Jupiter, more than three thousand; and from Saturn, five or six thousand.[2]

7.   That it is not possible to suppose the fixed Stars[3] [to be] too far away.[4]

As for the fixed Stars, the phenomena definitely prevent us from believing them to be closer to the Earth or the Sun than Saturn is. On the other hand, we do not observe anything to prevent us from supposing them to be separated from us by an extremely great distance.[5] And we can conclude, from what I shall say further on about the movements of the heavens, that they are so far away from the Earth that, in comparison with them, Saturn is extremely close.

[2] In the Copernican system, the planetary distances can all be calculated on the basis of the distance between the Earth and sun. The distances given here would be correct if the value for the distance to the sun given in Article 5 were accurate. In fact, the actual distance is about twenty times that calculated by Ancient astronomers and accepted in Descartes's time.

[3] The Latin term used here is 'Fixae', and it and the term for 'star' are usually capitalized when Descartes is discussing observable stars and normally not capitalized when he is discussing the nature of stars in general.

[4] The French title is: "That one may suppose the fixed Stars to be as far away as one pleases."

[5] One of Tycho Brahe's principal objections to the Copernican view arose from the fact that if the Earth has an annual revolution around the sun, observations made six months apart should reveal an apparent shift in the position of a star relative to the Earth. Copernicus had anticipated this and similar objections by claiming (correctly) that the stars were much more remote than had been previously believed, and thus that the effects of the Earth's revolution were too small to be measured. This required the distance to the nearest star to be at least 60 billion miles by the most conservative estimate. A Copernican who was prepared to accept so vast an enlargement of the universe (Saturn is only 886 million miles from the sun), might well then find the step to an infinite universe much easier. See Articles 40 and 41.

8.   That the Earth, if seen from the heaven, would only seem like a
            Planet smaller than Jupiter or Saturn.

It is obvious from the preceding that the Moon and the Earth would
appear much smaller, if seen from Jupiter or Saturn, than Jupiter or Saturn
appears if viewed from the Earth. Further, if the Sun were observed from
some fixed Star, it might not appear to be any bigger than the fixed Stars
appear when observed from our vantage point. Thus, if we wish to compare
the parts of the visible world with one another and judge their size without
prejudice; we must not believe the Moon, the Earth, or the Sun to be larger
than the Stars.

9.   That the Sun and the fixed Stars shine by their own light.

But, in addition to the fact that the Stars are not equal in size, we notice
another difference also: some shine by their own light, while others only
reflect the light which they receive from elsewhere. First of all, it cannot be
doubted that the Sun itself possesses the light with which it dazzles our eyes,
and indeed as much could not be obtained from all the fixed Stars taken
together; for they do not send us as much, although they are closer to us
than to the Sun. Further, if there were in the world some other more
brilliant body from which the Sun was borrowing its light, we would
necessarily have to see it. But if we consider also how bright and glittering
the rays of the fixed Stars are, despite the fact that they are extremely
distant from us and from the Sun, we will have no difficulty in believing that
they are like the Sun. Thus, if we were as close to one of them as we are to
the Sun, that Star would likely appear as big and luminous as the Sun.

10.   That the Moon and the other Planets derive their light from the
            Sun.

On the contrary, from the fact that the Moon shines only on the side
facing the Sun, we must conclude that it has no light of its own and merely
reflects toward our eyes the rays which it has received from the Sun. The use
of a telescope {recently} revealed the same thing to be true of Venus.[6] And
we can make similar judgments about Mercury, Mars, Jupiter, and
Saturn; because their light appears much weaker and less brilliant than that

---

[6] See the note to Article 16.

of the Fixed Stars,[7] and because they are not too far from the Sun to be illuminated by it.

11.   That the Earth does not differ from the Planets in the matter of light.

Finally, we know by experience that the same is true of the Earth; for it is composed of opaque bodies which reflect the Sun's rays no less strongly than the Moon does. Indeed, it is surrounded by clouds, which of course are much less opaque than most of its other parts. Nevertheless, when clouds are illuminated by the Sun, they often appear to us no less white than the Moon. Thus it is sufficiently clear that, in the matter of its light, the Earth does not differ from the Moon, Venus, Mercury, and the other Planets.[8]

12.   That the Moon, when new, is illuminated by the Earth.

We shall be even more assured of this if we take note of a certain weak light which appears on that part of the new Moon not illuminated by the Sun, and which we easily conjecture reaches it from the Earth by reflection; since it gradually diminishes as the part of the Earth which is illuminated by the Sun turns away from the Moon.

13.   That the Sun can be numbered among the fixed Stars and the Earth among the Planets.

So that, supposing we were on Jupiter and looking at our Earth, it is evident that it would appear smaller to us than Jupiter appears from here, but perhaps no less luminous. And the Earth would appear larger to us if we were on some closer Planet; but we would not see it at all if we were on one of the fixed Stars, because the distance would be too great. Thus, the Earth can be numbered among the Planets, and the Sun among the fixed Stars.[9]

---

[7] Further, unlike the stars, the apparent brightness of the planets varies considerably with their position relative to the Earth and sun.

[8] This is at least a first step in showing that the Earth *is* a planet.

[9] The categories being employed here are completely Copernican. 'Planet' comes from a Greek term meaning 'wanderer'; and in any but the Copernican system, the sun is a planet, since it appears to move relative to the stars. To argue that the sun is a fixed star because it, like them, is large and luminous and that the Earth is a planet because it is small and non-luminous is to have already rejected the traditional basis for the distinction between stars and planets; cf. Article 14.

14.   That the fixed Stars always remain in the same position, relative
             to one another, but that the same is not true of the Planets.

There is still another difference between the Stars: while some retain the
same position in relation to one another and remain the same distance
apart, which is why we call them fixed Stars; others are constantly changing
position, which is why we call them Planets or wandering stars.

15.   That various hypotheses may be used to explain the phenomena
             of the Planets.

Just as a man at sea in calm weather and looking at several other fairly
distant vessels, which seem to him to be changing position, is frequently
unable to say whether the change is caused by the movement of the vessel
on which he is or by that of the other vessels; when, from our situation, we
observe the course of the Planets and their various positions, even careful
observation does not always bring sufficient understanding to enable us to
determine, {from what we see}, to which bodies we ought properly to
attribute {the cause of} these changes. And since these changes are very
unequal and complicated, it is not easy to explain them, unless we choose
one of the various ways in which they can be understood, in accordance
with which we then suppose these changes to occur. To this end,
Astronomers have devised three different hypotheses or suppositions;
which they have merely attempted to make capable of explaining all the
phenomena, without considering whether they conformed to the truth.[10]

16.   That Ptolemy's hypothesis is not in conformity with
             appearances.

Ptolemy devised the first of these; but, as it is already commonly rejected
by all Philosophers, because it is contrary to several {recent} observations
(especially to the changes in light, similar to those which occur on the

---

[10] This is, of course, not true, it may refer to Osfander's unapproved preface to Copernicus' *De
Revolutionibus*... in which it is suggested that the Earth's motion be treated as a fiction, useful
for calculating planetary positions. Ptolemy, Copernicus, and Tycho, however, each insisted
that his system did conform to the truth.

Moon, which we observe on Venus[11]), I shall not speak further of it here.

17.   That those of Copernicus and Tycho do not differ if considered
             only as hypotheses.

The second is that of Copernicus and the third that of Tycho Brahe;[12]
considered purely as hypotheses, these two explain the phenomena equally
well, and there is not much difference between them.[13] Nevertheless, that of
Copernicus is somewhat simpler and clearer;[14] so that Tycho's only reason
for altering it was that he was attempting not merely a hypothetical
explanation but an account of how [he thought] this matter really was.

18.   That Tycho in words attributes less motion to the Earth than
             does Copernicus, but that in fact he attributes more.

Seeing that Copernicus had not hesitated to attribute motion to the
Earth; Tycho, to whom this opinion seemed not only absurd in Physics
but contrary to the common sense of men, tried to correct it. However,
because he did not give sufficient consideration to the true nature of

---

[11] The discovery that Venus had phases was announced by Galileo in 1610. In the Ptolemaic
system, see Plate III, Venus is always, roughly, between the sun and the Earth, since it never
appears more than 48 degrees from a line between the Earth and sun. Thus, an observer on the
Earth should never see more than a small crescent of Venus illuminated. Through the
telescope, however, Venus sometimes appears as a large crescent and sometimes as a much
smaller almost circular disk. The effect was predicted by Copernicus and shows that Venus, at
least, must orbit the sun. Also, through the telescope, the change in apparent size, and thus in
their distance from us, of Venus, Mercury, and Mars is far greater than the Ptolemaic system
can explain.

[12] See Plate V.

[13] In fact, there is almost no difference; both schemes will produce exactly the same relative
planetary and solar positions. Tycho's system is simply that of Copernicus with the Earth
taken as the unmoving reference point. Thus, while the moon and sun revolve around the
Earth, the other planets all revolve around the sun and are carried along with it. The only
observational differences would involve such things as stellar parallax; see note 5.

[14] Since the Tychonic system has all the advantages of the Copernican with none of the
apparent disadvantages, it may seem strange that Descartes did not favor it. It was, after all,
the more or less official view of the Church at the time and was compatible with Descartes's
own views on the relativity of motion. A glance at Plate V, however, will reveal that it is
difficult or impossible to imagine any *mechanism* which would produce such motion. For one
thing, the orbits of the sun and Mars interpenetrate; an impossibility if one believed, as
Descartes did, that something physical was moving the planets.

motion, he asserted only verbally that the Earth was at rest and in fact granted it more motion than had his predecessor.[15]

19. That I deny the motion of the Earth more carefully than Copernicus and more truthfully than Tycho.

That is why, although I do not differ at all from these two except on this one point, I shall deny the movement of the Earth more carefully than Copernicus and more truly than Tycho. I shall set forth here the hypothesis which seems to me the simplest and most useful of all; both for understanding the phenomena and for enquiring into their natural causes. And yet I give warning that I do not intend it to be accepted as entirely in conformity with the truth, but only as an hypothesis {or supposition which may be false}.

20. That we must suppose the fixed Stars to be a very great distance from Saturn.

First, because we do not yet know with certainty what distance separates the fixed Stars from the Earth, and because it is impossible to imagine them so far away that this is contrary to the phenomena, let us not be content to merely place them beyond Saturn, where all {Astronomers} agree that they are, but let us take the liberty of supposing them to be {as} far beyond Saturn {as will serve our purpose}. For if we tried to judge their altitude by comparison with the distances between the bodies which we see on the earth, that which is already conceded to them would be as unbelievable as a distance very much greater. If, on the other hand, we consider the omnipotence of God who created them, the greatest distance of which we can conceive is no less credible than a smaller one. And I shall show further on that neither the phenomena of the Planets nor indeed those of the Comets can be satisfactorily explained unless we suppose that there is a very great distance between the fixed Stars and the sphere of Saturn.

21. That the substance of the Sun, like that of fire, is extremely mobile; but that this does not cause it to move {as a whole} from one place to another.

Secondly, since the Sun gives off its own light, like fire and the fixed Stars; let us suppose that it resembles fire in its motion and the fixed Stars in its

---

[15] See Articles 28, 38, and 39 for Descartes's defense of this claim.

situation. There is certainly nothing more mobile than fire to be seen on the earth, since, if the bodies it touches are not thoroughly hard and solid, it gradually disintegrates them {and disunites all their particles} and moves them with it. However, its motion consists only in each of its parts moving in relation to the others, for the fire as a whole does not move from one place to another unless it is transported by some body to which it is adhering. Similarly, we can judge that the Sun is composed of very fluid and mobile matter, which carries the surrounding parts of the heaven along with it. Nevertheless, we can judge that it resembles the fixed Stars in that it does not move from one place in the heaven to another.

22.   That the Sun differs from fire in that it does not need fuel.

And there is no reason to find fault with my comparison of the Sun to fire on the grounds that all fire that we see here needs fuel, and we do not observe the same [to be true] of the Sun. According to the laws of nature, fire (like any other body), having once been formed, always continues to exist unless destroyed by some external cause. However, since it is composed of the most fluid and mobile matter possible, here on earth it is constantly being dissipated by the matter which surrounds it. Thus, fire does not, strictly speaking, need fuel in order to be maintained exactly as it is, but only so that another new fire may replace the original as it is extinguished. However, the Sun is not similarly dissipated by the heavenly matter which surrounds it; therefore, we have no reason to judge that it needs to be fed like fire. {I hope to show} later[16] that {the Sun also resembles fire in that} new matter constantly enters it while other matter leaves it.

23.   That the fixed Stars do not all turn on the same sphere, but that each one has a vast space around it, empty of other fixed Stars.

In addition, we must notice here that, while the Sun and the fixed Stars resemble each other as far as their situation is concerned, they are not all situated on the circumference of a single sphere, as some suppose; because the Sun cannot be with them on that circumference. Rather, just as the Sun is surrounded by a vast space in which there is no fixed Star, so also each fixed Star must be very distant from all others, and some of these Stars must be more distant from us and from the Sun than others are. So that, if S, for

[16] Article 69.

example, is the Sun,[17] F, f will be fixed Stars, and we will understand that numerous others exist, above, below, and beyond the plane of this figure, scattered throughout all the dimensions of space.[18]

24.   That the heavens are fluid.

Third, it must be thought that the matter of the heaven, like that which forms the Sun and the fixed Stars, is fluid. This is an opinion which is now commonly held by all Astronomers, because they see that otherwise it is almost impossible to give a satisfactory explanation of the phenomena of the Planets.[19]

25.   That the heavens carry with them all the bodies which they contain.

But it seems to me that many who seek to attribute to the heaven the property of fluidity are mistaken in imagining it to be an entirely empty space, which not only offers no resistance to the motion of other bodies, but also lacks the force to {move them and} carry them along with it. For in addition to the fact that such a void cannot exist in nature, there is a factor which all fluids have in common: the reason that they do not offer so much resistance to the motions of other bodies is {not that they contain less matter, but} that they also have motion {of their particles} in themselves. And since this motion can be easily determined in any direction, if it has been determined in some single direction, then a fluid will necessarily, by the force of this motion, carry with it all the bodies which it contains and which are not prevented from following it by some external cause, even though these bodies may be entirely at rest, and hard and solid; as manifestly follows from what has been said above[20] {about the nature of fluid bodies}.

[17] See Plate VI.

[18] Previously, it had usually been assumed, even by Copernicus, that the stars were all about the same distance from the center of the universe. With the rejection of the crystalline spheres, however, the assumption becomes unnecessary.

[19] Presumably, Descartes has in mind here the fact that comets move through the heavens.

[20] Part II, Article 61.

26.   That the Earth is at rest in its heaven which nevertheless carries
              it along.

Fourth, since we see that the Earth is not supported by columns or
suspended in the air by means of cables but is surrounded on all sides by a
very fluid heaven, let us assume that it is at rest and has no innate
tendency to motion, since we see no such propensity in it. However, we
must not at the same time assume that this prevents it from being carried
along by {the current of} that heaven or from following the motion of the
heaven without however moving itself: in the same way as a vessel, which is
neither driven by the wind or by oars, nor restrained by anchors, remains at
rest in the middle of the ocean; although it may perhaps be imperceptibly
carried along by {the ebb and flow of} this great mass of water.

27.   That the same is to be believed of all the Planets.

And just as the other Planets resemble the Earth in being opaque and
reflecting the rays of the Sun, we have reason to believe that they also
resemble it in remaining at rest, each in its own part of the heaven, and that
the variation we observe in their position results solely from the motion of
the matter of the heaven which contains them.

28.   That the Earth, properly speaking, is not moved, nor are any of
              the Planets; although they are carried along by the heaven.

And it is important to remember here what was said earlier concerning
the nature of movement; i.e., that (if we are speaking properly and in
accordance with the truth of the matter) it is only the transference of a body
from the vicinity of those bodies which are immediately contiguous to it,
and considered to be at rest, into the vicinity of others. However, in
common usage, all action by which any body travels from one place to
another is often also called movement; and in this sense of the term it can
be said that the same thing is simultaneously moved and not moved,
according to the way we diversely determine its location. From this it
follows that no movement, in the strict sense, is found in the Earth or even
in the other Planets; because they are not transported from the vicinity of
the parts of the heaven immediately contiguous to them, inasmuch as we
consider these parts of the heaven to be at rest. For, to be thus transported,
they would have to be simultaneously separated from all {the contiguous

parts of the heaven}, which does not happen. However, because the matter of the heaven is fluid, sometimes some of its particles, and sometimes others, move away from the Planet to which they are contiguous, and this by a movement which must be attributed solely to them and not to the Planet: in the same way as the partial transferences of water and air which occur on the earth's surface are usually attributed, not to the earth, but to those portions of water and air which are transported.

29.   And that no motion is to be attributed to the Earth, even if we use 'motion' improperly, according to common usage; but that it would then correctly be said that the other Planets are moved.

And if one takes 'motion' in the popular sense, one can quite well say that all the other Planets are moved, even the Sun and the fixed Stars; but it is very improper to speak of the Earth in this way. For the common people determine the location of the Stars by certain points on the Earth which they consider to be motionless, and believe that the Stars are moved when they leave the locations thus determined: this is convenient for everyday life and therefore reasonable. Indeed, in our childhood, we all believed that the Earth was flat and not spherical; and that the top, the bottom, and the four cardinal points of the world, namely, the North, the South, the East, and the West; were always and everywhere the same. We accordingly indicated by means of those points, {which are fixed only in our minds}, the locations of other bodies. But if a Philosopher, ({professing to search for truth and} having observed that the Earth is a globe floating in a fluid heaven whose parts are extremely mobile and that the fixed Stars always maintain the same position in relation to one another) were to consider these Stars as motionless and attempt to use them to determine the location of the Earth, and were to conclude from this that the Earth moves; he would be {in error and} speaking in a manner contrary to reason. For 'location' in its {true and} philosophical sense must be determined by the bodies immediately contiguous to that which is said to be moved, and not by those which are extremely distant; as are the fixed Stars {in relation to the Earth}.[21] And if one interprets it according to common usage, one has no reason to believe that the Stars, rather than the Earth, are motionless (unless one imagines that there are no other bodies beyond the Stars from which the Stars can be separated and in relation to which one could say that they move and the

---

[21] Of course, the question at issue was exactly that of the Earth's motion relative to the fixed stars.

Earth remains at rest; in the same sense as one claims to be able to say that the Earth moves in relation to the fixed Stars). But this is contrary to reason, since the nature of our intellect is such that it perceives no limits to the universe and since, consequently, anyone who takes careful notice of the greatness of God and the weakness of our perception will judge that it is much more appropriate to believe that perhaps, beyond all the fixed Stars which we see, there are other bodies in relation to which we would have to say that the Earth is at rest and all the Stars move together, than to suppose {the Creator's power so imperfect} that none such could exist, {as must be the belief of those who state in this way that the Earth moves. However, if, in spite of this, conforming to common usage, we seem further on to attribute some motion to the Earth, it will have to be remembered that we are speaking improperly, in the way in which it is sometimes possible to say, of passengers who lie sleeping in a ship, that they nevertheless go from Calais to Dover, because the vessel takes them there}.

30.   That all the Planets are carried around the Sun by the heaven.

Now that we have, by this reasoning, removed any possible doubts about the motion of the Earth, let us assume that the matter of the heaven, in which the Planets are situated, unceasingly revolves, like a vortex having the Sun as its center, and that those of its parts which are close to the Sun move more quickly than those further away; and that all the Planets (among which we {shall from now on} include the Earth) always remain suspended among the same parts of this heavenly matter. For by that alone, and without any other devices, all their phenomena are very easily understood. Thus, if some straws {or other light bodies} are floating in the eddy of a river, where the water doubles back on itself and forms a vortex as it swirls; we can see that it carries them along and makes them move in circles with it. Further, we can often see that some of these straws rotate about their own centers, and that those which are closer to the center of the vortex which contains them complete their circle more rapidly than those which are further away from it. Finally, we see that, although these whirlpools always attempt a circular motion, they practically never describe perfect circles, but sometimes become too great in width or in length, {so that all the parts of the circumference which they describe are not equidistant from the center}. Thus we can easily imagine that all the same things happen to the Planets; and this is all we need to explain all their remaining phenomena.

31.   How the individual Planets are carried along.

Let us then suppose that S[22] is the Sun, and that all the surrounding matter of the heaven turns in the same direction, namely from the West to the East via the South, or from A to C via B, assuming the North Pole to be elevated above the plane of this figure. As a result, the matter which surrounds Saturn takes almost thirty years to carry it completely around the circle marked ♄; and that the matter which surrounds Jupiter carries it, {together with the other little Planets which accompany it,[23] all the way} around the circle ♃ in twelve years. By the same means, Mars in two years, the Earth and the Moon in one year, Venus in eight months, and Mercury in three, complete the revolutions which are indicated by the circles marked [respectively] ♂, T, ♀, ☿.

32.   How the spots {which are seen on the surface} of the Sun are transported.

{Let us also suppose that} those opaque bodies which are visible on the surface of the Sun with the aid of a telescope, and which we call its spots; lie on its surface and take twenty-six days to complete their revolution.[24]

33.   How the Earth is also moved around its own center and the Moon around the Earth.

In addition, in the great vortex which forms a heaven[25] {having the Sun at its center}, there are other smaller ones which we can compare to those I have often seen in eddies of rivers where they {all follow the current of the larger vortex which carries them, and} move in the direction in which it moves. One of these vortices has Jupiter at its center, and moves with it the four satellites which revolve around Jupiter, at speeds so proportioned that the most distant of the four completes its revolution in about sixteen days, the next one in seven, the third in eighty-five hours and that closest to the center in forty-two hours; and thus, they revolve several times around

[22] See Plate VII.
[23] The reference is to the four major satellites of Jupiter, discovered by Galileo in 1609.
[24] There was considerable debate at the time as to whether sunspots were on the surface of the sun or in orbits very close to the surface, like clouds.
[25] By 'heaven', Descartes means a large, spherical mass of rotating fluid material, having a sun at its center. Thus, there are as many heavens as there are stars.

Jupiter while it describes a large circle around the Sun. Similarly, the vortex which has the Earth at its center carries the Moon around the Earth in the space of a month, while the Earth turns on its own axis in the space of twenty-four hours; and in the time it takes the Earth and the Moon to describe the circle which is common to them {and which forms the year}, the Earth revolves on its axis {approximately} three hundred and sixty-five times and the Moon revolves {approximately} twelve times around the Earth.

34.   That the movements of the heavens are not perfectly circular.

Finally, we must not think that all the centers of the Planets are always situated exactly on the same plane, or that the circles they describe are absolutely perfect; let us instead judge that, as we see occurring in all other natural things, they are only approximately so, and also that they are continuously changed by the passing of the ages.

35.   Concerning the variations of the Planets in latitude.

Now, if this figure[26] represents the plane in which the center of the Earth is situated, which is called the plane of the Ecliptic[27] and determined by means of the fixed Stars; each of the other Planets must be thought to revolve in other planes, each slightly inclined to the Ecliptic plane and intersecting it in a line which passes through the center of the Sun; so that the Sun is found in all these planes.[28] For example, the orbit of Saturn now intersects the Ecliptic at the signs of Cancer and Capricorn, but is above that plane, inclined toward the North, at Libra, and inclined toward the South at Aries; and the angle of its inclination is about two and a half degrees. Similarly, the other Planets revolve in planes which intersect that of the Ecliptic in other lines. The inclination of the planes of Jupiter and Mars [to the ecliptic] is smaller than that of Saturn, that of Venus larger by about one degree, and that of Mercury is greatest, measuring about seven degrees. Moreover, the spots which appear on the Sun's surface also revolve around it in planes inclined to that of the Ecliptic at seven degrees

[26] Plate VII.
[27] The ecliptic is the apparent annual path of the sun around the zodiac. Since it is, in effect, the projection against the stars of the Earth's annual revolution; the plane of the ecliptic is simply the plane of the Earth's orbit.
[28] This was first announced by Kepler in his *Mysterium Cosmographicum*, 1596.

or more (at least if the observations of Father Scheiner, S. J.,[29] are correct, and he made them so carefully that there seems no reason to desire any more). Thus, in this respect, their motion does not differ from those of the Planets. The Moon also is transported around the Earth on a plane inclined at five degrees to that of the Ecliptic. Finally, the Earth itself rotates around its center on the plane of the Equator, which it always carries with it, and which is inclined to the Ecliptic by twenty-three and one-half degrees. And the deviations {in the number of degrees} between the Ecliptic and the location of a Planet are called the Planet's latitudinal movements.

36.   Concerning their longitudinal movement.[30]

However, their revolutions about the Sun are called their longitudinal movements. And these movements also deviate in that, {since} they are not always equidistant from the Sun, {they do not always seem to move at the same speed in relation to it}.[31] For, in this age, Saturn is more distant from the Sun, by about one-twentieth of the distance {between them}, when it is in {the sign of} Sagittarius, than when it is in {the sign of} Gemini; when Jupiter is in Libra, it is further away than when it is in Aries; and thus also the other Planets have their Aphelia and Perihelia in other places. But, a few centuries from now, all these things will be observed to have changed {from the way in which they now are}, and {those who will be living at that

[29] In 1611, Christopher Scheiner, a Jesuit lecturer at the University of Ingoldstadt, announced that he had discovered spots on the sun. A long and bitter controversy ensued between him and Galileo concerning the priority of the discovery and the nature of the spots themselves. In 1630, Scheiner published *Rosa Ursina* . . ., which contained his observations and views on the nature of the spots and the rotation of the sun. Concerning this book, Descartes wrote to Mersenne in 1634: "However, the observations which are in this book provide so many proofs to contradict the movement attributed to the Sun [its annual motion around the ecliptic], that I cannot believe that Father Scheiner does not, in his own mind, accept Copernicus' opinion; . . . .": A. & T., I, 281–282.

[30] The longitude of a celestial body is measured in degrees along the ecliptic, beginning at the point of the vernal equinox, which is one of the two points at which the apparent path of the sun intersects the celestial equator. Since that point moves, see note 34, the longitude of even a fixed star thereby changes.

[31] This additional claim is, of course, true. In accordance with Kepler's second law of planetary motion, planets move fastest when they are closest to the sun. The remainder of the text, however, seems to indicate that this is not what Descartes had in mind but that he was considering the longitude of the aphelia and perihelia (terms introduced by Kepler) of the planets. Further, there does not seem to be any evidence that Descartes knew Kepler's laws. He certainly never attempts to deduce them from his assumptions (as Newton was able to do from his), and seems acquainted only with Kepler's work in optics.

time will be able to observe that} the individual Planets, and also the Earth, will intersect the plane on which the Ecliptic now is at different places {from those at which they now intersect it},[32] and will be slightly more, or slightly less, inclined to the Ecliptic, and will not be found in the same signs when they are at the points at which they are closest to, or farthest from, the Sun.[33]

37.   That all the phenomena of the Planets can be explained by the hypothesis proposed here.

It is not necessary for me to go on from this to explain how we can understand, from this hypothesis, the phenomena of day and night, of summer and winter, or of the Sun's approaching the Tropics and receding, of the phases of the Moon, of Eclipses, of the stations and retrogressions of the Planets, of the precession of the equinoxes,[34] of the variation which we observe in the obliquity of the Ecliptic,[35] and of other similar things: for all this is easy for those who have some knowledge of Astronomy.

[32] Of course, at any given time, the planes of the Earth's orbit and of the ecliptic do not intersect; they are one and the same. This remark is doubtless simply the result of a syntactic confusion on Descartes's part.

[33] Descartes seems to be making a number of claims here which would have been quite surprising to his contemporaries. First, he seems to claim that the aphelion and perihelion points of all the planets, and also their points of intersection with the ecliptic, are changing relative to the stars. The second claim is that the inclination of the planetary orbits is changing as well. However, the fact that he locates the aphelion and perihelion points with reference to the signs of the zodiac, plus the syntactic confusion mentioned in the previous note, makes it likely that he has something much less startling in mind. If one takes him to be considering the future inclination to the present ecliptic in the case of the Earth's orbit and to the future altered ecliptic in the case of the other planetary orbits, then all the changes he mentions would occur simply as a result of variation in the obliquity and precession of the equinoxes. See notes 34 and 35.

[34] The precession of the equinoxes was discovered by Hipparchus about 125 B.C. It results from the fact that the Earth's poles complete a rotation in a circle of 47 degrees about once every 26,000 years. The result is that the equinoctial points move slowly along the ecliptic. Thus, the longitude and latitude of celestial bodies, which are measured from the vernal equinox, constantly change. Further, the signs of the zodiac, which are marked off in twelve equal divisions of the band of the zodiac from the vernal equinox, move with it. Thus, the original constellations are no longer located in the signs which were named for them, and the signs move slowly relative to the fixed stars.

[35] The obliquity of the ecliptic is the angle between the ecliptic and the celestial equator. Copernicus believed that the obliquity varied between 24 and $23\frac{1}{2}$ degrees in a period of 3,500 years. This seems to have resulted from his acceptance of inaccurate determinations of the

38.   That, according to Tycho's hypothesis, the Earth ought to be
           said to revolve around its center.

However I shall indicate briefly here how Brahe's hypothesis, which is
commonly accepted by those who reject that of Copernicus, attributes
more motion to the Earth than the latter does. First, while the Earth, in
Tycho's opinion, remains motionless, the heaven with its stars must
revolve around it each day, which is inconceivable unless we also think that
all parts of the Earth are transferred from all the parts of the heaven which
they were touching a short while before, and that they proceed to touch
others. Now because this transference is reciprocal, as has already been
stated,[36] and because there must be as much force or action in the Earth as
in the heaven; this transference gives us no reason to attribute motion to
the heaven rather than to the Earth. Moreover, in accordance with what
was said earlier, this motion should be attributed only to the Earth;
because the separation takes place over its whole surface, and does not
similarly occur over the whole surface of the heaven but only in the concave
portion contiguous to the Earth; which is an extremely small area in
comparison to the convex part. And it makes no difference that they say
that, in their opinion, the convex surface of the starry heaven also
undergoes a separation from the other surrounding heaven, namely the
crystalline or Empyreal, in the same way as the concave surface of the same
heaven is separated from the earth,[37] and that this is why they attribute the
motion to the heaven rather than to the earth. For they have no proof to

value by his predecessors. Further, the very accurate determination by Tycho gave a slightly
larger value than that given by Copernicus about sixty years previously. Thus, it was
commonly believed that the obliquity fluctuated fairly rapidly. Curiously, the obliquity does
fluctuate, but the effect is too small to be reliably observed by the naked eye, even over a period
of 2,000 years.

A marginal note in what is known as the annotated copy of the *Principles* (annotator
unknown) states:

"That is to say, the variation which occurs in the declination of the Ecliptic in relation to the
Equator, which is now inclined at 23 and one half degrees. And in the time of Copernicus, this
was only 23° 24'. And in the time of Ptolemy, it was inclined at an angle of 23° 54'.* And that is
why the Astronomers had imagined a crystalline Heaven which swung irregularly, and very
slightly, from South to North, and from North to South, so that in our year of 1659, the
declination is gradually increasing."

* Ptolemy actually gives a value of 23° 51'; Copernicus gives "(very nearly) 23 28'." Cf. Article
156.

[36] Article 28, and Part II, Articles 28 and 29.

[37] There seems to be no reason why 'earth' is sometimes not capitalized in this article.

demonstrate this separation of the whole convex surface of the starry
heaven from the other surrounding heaven, but merely claim that it occurs.
Thus, according to their hypothesis, the reason for which we ought to
attribute motion to the heaven and immobility to the Earth is uncertain and
depends entirely on their imagination; while, on the other hand, the reason
for which they should say that the earth moves is obvious and certain.

39.   And that [the Earth] must also be said to be moved around the
           Sun by an annual movement.

According to this same hypothesis of Tycho's, the Sun, as it rotates
around the Earth in an annual movement, carries with it not only Mercury
and Venus, but also Mars, Jupiter and Saturn, which are further from it
than the Earth is. This cannot be asserted, especially of a heaven which is
fluid, as they suppose it to be; unless all the intermediate matter of the
heaven is simultaneously carried along, and unless the Earth is meanwhile
separated by some force from the parts of this matter which are contiguous
to it, and describes a circle relative to this material. Therefore, this
separation, which belongs to the whole Earth and requires an individual
action in it, will instead have to be called its movement.

40.   That a change in the Earth's position produces no diversity in
           the appearance of the fixed Stars, because of their extreme
           remoteness.

Here an objection to my hypothesis can be made, namely that, since the
Sun always retains the same position in relation to the fixed Stars, the
Earth, which revolves around the Sun, must move closer to these Stars and
then farther away from them while it is describing that great circle on its
year-long journey; and yet we have not been able to find this from the
observations which have been made.[38] However, {it is easy to reply that} the
great distance between the Earth and the fixed Stars explains this: for I
suppose this distance to be so immense that the whole circle which the
Earth describes around the Sun should be counted as a mere point in
comparison to it. This will perhaps seem unbelievable to those who have
not accustomed their minds to the contemplation of God's mighty works,
and who think that the Earth is the most important part of the universe,
because it is the dwelling of man, for whose benefit {they believe, without

[38] Stellar parallax was first successfully observed in 1838, by Bessel.

foundation, that} all things were created; but {I am sure that} Astronomers, who all already know that the Earth, in comparison with the heaven, takes up no more space than a point,[39] will not find it so strange.

41.[40]   That this distance of the fixed Stars is necessary for the motion of the Comets, which it is now certain are in the heaven.

Further, the Comets, which it is now sufficiently certain are not situated in our atmosphere, as overly-ignorant Antiquity believed {before Astronomers had investigated their parallaxes}, require this extremely vast space between the sphere of Saturn and the fixed Stars in order to complete all their journeys: for these are so varied, so immense, and so dissimilar to the stability of the fixed Stars and to the regular revolutions of the Planets around the Sun, that they would seem impossible to explain by means of any laws of nature without this space. And we must not be influenced by the fact that Tycho and the other Astronomers who carefully searched for the parallaxes of the Comets said only that they were situated beyond the Moon, toward the sphere of Venus or Mercury but not beyond Saturn; for they could no less satisfactorily have deduced from their calculations that the Comets were beyond Saturn. But because they were disputing the views of the ancients, who included the Comets among the meteors {formed in the air} below the Moon, they contented themselves with showing that they are in the heaven, and did not dare attribute to them all the altitude which their

---

[39] The horizon *appears* to exactly bisect the stellar sphere. Since the plane of the horizon is tangent to the spherical Earth, the Earth must be *much* smaller than the stellar sphere to account for the apparent bisection.

[40] The following is written opposite this article in the margin of the annotated copy: "From here on, the translation is by Mr. D." (MS note in one hand, perhaps that of Clerselier? What follows is in another hand, certainly that of Legrand): "We judge this to be the case because we have in our hands the original, written in his own hand [originally "written in Mr. Desc.'s own hand"; these words crossed out]. And it is unbelievable that, if this translation were not his own, he would have taken the trouble to copy it out, especially as he was so overburdened with business." The same claim occurs in a different copy of the *Principles*, and in a different hand; possibly that of Ozanam, a 17th-century Mathematician. Adam and Tannery reject the claim that Descartes himself was responsible for the remainder of the French translation, for what seem to be good reasons. However, the existence of the manuscript at least shows that Descartes was well acquainted with the French translation from this point on, and that he apparently had no serious objections to it: A. & T., IX-2, x-xvi.

calculations were revealing, for fear of making their proposition less
believable.[41]

42.   That everything which we see here on Earth is also relevant to
                 the phenomena, but that it is not necessary for us to consider
                 them all from the beginning.

In addition to these rather general things, I could also include here,
among the phenomena, several other specific things, concerning not only
the Sun, the Planets, the Comets, and the fixed Stars, but also the Earth:
that is, everything which we see {around the Earth, or which occurs} on its
surface. For indeed, in order to know the true nature of this visible world, it
is not sufficient to find some causes by which one can explain what appears
in the heaven very far from us; it is necessary also to be able to deduce from
them the things we see very close to us {and which affect us more
noticeably}. But, even so, {I think} it is unnecessary for us to consider them
all {immediately, and deem it preferable for us to attempt first} to discover
the causes of these more general things {which I have enumerated, so that
we may see subsequently whether we can also deduce from these same
causes all the more specific things which we shall have ignored while
searching for these causes}. And we shall know that we have correctly
determined these causes when we observe that we can explain, by their
means, not only those phenomena which we have considered up to now,
but also everything else about which we have not previously thought.[42]

43.   That it can scarcely be possible that the causes from which all
                 phenomena are clearly deduced are false.

And certainly, if the principles which I use are very obvious, if I deduce
nothing from them except by means of a Mathematical sequence, and if
what I thus deduce is in exact agreement with all natural phenomena; it

[41] In fact, comets pass fairly close to the sun, many passing closer to it than does the Earth.
They are generally not visible until they approach to about twice the Earth's distance from the
sun.
[42] In a letter to Morin, dated July 13, 1638, Descartes writes: "Finally, you say that there is
nothing so easy as adapting some cause to an effect. But although there certainly are several
effects to which it is easy to adapt diverse causes, one [cause] to each [effect], yet it is not so easy
to adapt one identical cause to several different effects, unless it is the true cause from which
they originate. There are even often some [effects] which are such that to give a cause from
which they can be clearly deduced is sufficient proof of their true cause.": A. & T., II, 197–199.

seems {to me} that it would be an injustice to God to believe that the causes of the effects which are in nature and which we have thus discovered are false. For we would then be accusing Him of having made us so imperfect as to be liable to make mistakes, even when correctly using our reason {which He has given us}.

44.   That I nevertheless wish those [causes] I am proposing here to be taken only as hypotheses.

However, {because the matters I am treating here are of no little importance,} and because I should perhaps be thought too presumptuous if I stated that I had discovered truths {which others have failed to discover}, I prefer to make no decision about it; and, {in order that each reader may be free to form his opinion}, I wish what I shall write later to be taken only as an hypothesis {which is perhaps very far from the truth}. But, even though these things may be thought to be false, I shall consider that I have achieved a great deal if all the things which are deduced from them are entirely in conformity with the phenomena: for, if this comes about, my hypothesis will be as useful to life as if it were true, {because we will be able to use it in the same way to dispose natural causes to produce the effects which we desire}.

45.   That I shall even assume here some which it is certain are false.

Indeed, in order to better explain natural things, I may even retrace their causes here to a stage earlier than any I think they ever passed through. {For example}, I do not doubt that the world was created in the beginning with all the perfection which it now possesses; so that the Sun, the Earth, the Moon, and the Stars existed in it, and so that the Earth did not only contain the seeds of plants but was covered by actual plants; and that Adam and Eve were not born as children but created as adults. The Christian faith teaches us this, and natural reason convinces us that this is true; because, taking into account the omnipotence of God, we must believe that everything He created was perfect in every way. But, nevertheless, just as for an understanding of the nature of plants or men it is better by far to consider how they can gradually grow from seeds than how they were created [entire] by God in the very beginning of the world; so, if we can devise some principles which are very simple and easy to know and by which we can demonstrate that the stars and the Earth, and indeed

everything which we perceive in this visible world, could have sprung forth
as if from certain seeds (even though we know that things did not happen
that way); we shall in that way explain their nature much better than if we
were merely to describe them as they are now, {or as we believe them to have
been created}.[43] And because I think I have discovered some principles of
this kind, I shall here briefly expound them.

46.   What these assumptions, which I am making here for the
             explanation of all phenomena, are.

We noticed earlier that it is certain that all the bodies which compose the
universe are formed of one [sort of] matter, which is divisible into all sorts
of parts and already divided into many which are moved diversely and the
motions of which are in some way circular, and that there is always an equal
quantity of these motions in the universe: but we have not been able to
determine in a similar way the size of the parts into which this matter is
divided, nor at what speed they move, nor what circles they describe. For,
seeing that these parts could have been regulated by God in an infinity of
diverse ways; experience alone should teach us which of all these ways He
chose. That is why we are now at liberty to assume anything we please,
provided that everything we shall deduce from it is {entirely} in conformity
with experience. Let us therefore suppose, if you please, that God, in the
beginning, divided all the matter of which He formed the visible world into
parts as equal as possible and of medium size, that is to say that their size
was the average of all the various sizes of the parts which now compose the
heavens and the stars. And let us suppose that He endowed them
collectively with exactly that amount of motion which is still in the world at
present. And, finally, that He caused them all to begin to move with equal
force {in two different ways, that is}, each one separately around its own
center, by which means they formed a fluid body, such as I judge the heaven
to be; and also several together around certain other centers equidistant
from each other, arranged in the universe as we see the centers of the fixed
Stars to be now; and also around other somewhat more numerous

---

[43] Descartes's view here seems to rest on the assumption that in order for things to obey the
laws of nature. God must have created them as the kinds of things which could have been
produced by those laws alone. See Part IV, Article 1.

points,[44] equal in number to the Planets {and the Comets}. Thus, for example, {we may think that God divided} all the matter which is in the space AEI[45] {into a very great number of particles which} He transported {not only each around its center but also} all together around the center S; and that, similarly, He transported all the parts of the matter which occupies the space AEV around the center F, and so on; so that [these] parts formed as many vortices {from now on, I shall use this word to denote all the matter which thus revolves around each of these centers} as there are now heavenly bodies in the world.

47.   That their falseness does not prevent what will be deduced from them from being true and certain.

These few {suppositions} seem to me sufficient for all the effects of this world to result from them in accordance with the laws of nature explained previously, as if they were [the] causes [of these effects]. And I do not think it possible to devise any simpler, more intelligible or more probable principles than these. For, although these laws of nature are such that, even if we were to assume [the existence of] the Chaos {of the poets, that is, a total confusion of all parts of the universe}; we could still demonstrate that, by these laws, this confusion must {gradually} be transformed into the order which is at present in the world. And although I formerly undertook to explain how this could have happened;[46] however, because {to make Him the author of} confusion seems to be less in agreement with the supreme perfection of God the creator of things than [does] proportion or order, and is also less distinctly perceived by us, and because no proportion and no order is simpler or easier to know than that which consists in equality of all kinds: I am accordingly supposing here that all particles of matter were, in the beginning, equal to one another both in size and motion; and I leave no inequality in the universe except for that which is in the situation of the fixed Stars, and which appears so clearly to anyone who gazes upon the

---

[44] The Latin expression used here is 'aliquanto plura', which, in Classical Latin, would be translated as it is here: indicating that Descartes believed that there were more planets and comets than there are stars. In Le Monde, Descartes had described God as creating a universe in which each star possessed its own planetary system, although he emphasized again that this was an imaginary process, used as an aid to the understanding. In Descartes's time, the expression could mean 'sufficiently numerous'; which would make the passage read, "...other points sufficient in number to equal the number of Planets...."

[45] See Plate VI.

[46] See Discourse on Method, V, and Le Monde, VI-X.

heaven at night that it plainly cannot be denied. Besides, the way in which I assume this matter to have been arranged {in the beginning} is of very little importance, since this arrangement must subsequently have been changed, in accordance with the laws of nature. And it is almost impossible to imagine any arrangement from which we could not deduce, by these laws, the same effect; {since it must change continually, until it finally forms a world exactly similar to this one} (although perhaps with more difficulty {from some suppositions than from others}). Because, given that these laws cause matter to assume successively all the forms it is capable of assuming, if we consider these causes in order, we shall finally be able to reach the form which is {at present} that of this world. {I state this here deliberately, so that it may be noticed that, although I speak of suppositions, I nevertheless do not make any which, even if known to be false, could give rise to doubts about the truth of the conclusions which will be drawn from them.}[47]

48.   How all the particles of the heavenly matter became spherical.

And so, in order that we may begin to show the efficacy of the laws of nature in the proposed hypothesis, let us consider that, since all the matter of which the world is composed was in the beginning divided into many equal parts, these could not at first have been spherical; for several spheres joined together do not {form a completely solid and continuous body,[48] like this universe, in which, as I demonstrated earlier, there can be no void}. However, no matter what shape these parts may have had at that time, it was impossible for them not to become spherical with the passing of time because of their various circular motions. And because the force by which they were moved in the beginning was sufficient to separate them from one another; that same force, enduring {in them subsequently}, was also undoubtedly great enough to break off all their angles as they came in contact with one another, for this effect required less force than the previous one had. And solely from the fact that all the angles of a body are thus worn down, we easily understand that each at length became spherical: because every part of that body which protrudes beyond the spherical figure is here referred to as an angle.

[47] The Latin simply says, "In this way no error from false supposition need be feared here."
[48] The Latin simply says, "...do not fill a continuous space."

49.   That around these spherical particles there must be other more subtle matter {to fill all the space in which they are situated}.

However, inasmuch as there cannot be any empty space anywhere {in the universe}, and because the parts of matter, being spherical, cannot unite closely enough to avoid leaving certain little intervals {or spaces} around themselves: these spaces must be filled by certain other scrapings of matter which must be extremely tiny and able to change their shapes at any moment in order to conform to those of the places they enter. For in fact, that which is detached from the angles of those particles of matter which are becoming spherical, by being gradually worn down, is so tiny, and acquires such great speed, that the sole force of its motion can divide it into innumerable scrapings which, {being of no determined size or shape, easily} fill all the angles {or spaces} into which the other parts of matter cannot enter.

50.   That the particles of this more subtle matter are very easily divided.

For it must be noted that the smaller these scrapings of other particles are, the more easily they can be moved and {subsequently} reduced to parts even tinier: because the smaller they are, the more surface area they have in proportion to their bulk, {and the size of this surface area causes them to meet correspondingly more bodies which attempt to move or divide them, while their small quantity of matter makes them correspondingly less able to resist their force}: and they encounter bodies in proportion to their surface and are divided according to their bulk.

51.   And that these particles move very rapidly

We must also notice that, although the scrapings {thus detached from the parts of matter which are becoming rounded} have no motion which does not come from these parts, they must however move much more quickly. The reason is that the [larger] parts, which travel by straight and open paths, drive this scraping {or dust which is among them} through other paths which are oblique and narrower. Similarly, we see that by closing a bellows quite slowly, we force the air out of it quite rapidly, because the opening through which this air emerges is small. And it has already been shown above that there must {necessarily} be some part of matter which moves extremely quickly and is divided into {an} indefinite {number of} particles, so that {all the} circular and dissimilar motions {which are in the world} can occur without any rarefaction or void; and {I think that} no other [kind of matter] {more} suited to that effect can be found {or imagined}.

52.   That there are three elements of this visible world.

And so now we have two very different kinds of matter, which can be called the two first elements of this visible world. The first is that of the matter which has so much force of agitation that, by colliding with other bodies, it is divided into particles of indefinite smallness, and which adapts its shapes to fill all the narrow parts of the little angles left by the others. The second is that of the matter which is divided into spherical particles, admittedly very small if compared with those bodies which our eyes can discern; yet of a certain and determined quantity and divisible into others much smaller. And, in a short while, we shall discover the third, which is composed of parts which are either much bulkier or have shapes less suited to movement. And we shall show that all the bodies of this visible world are composed of these three elements: the Sun and the fixed Stars of the first, the Heavens[49] of the second, and the Earth, the Planets, and the Comets of the third. For since the Sun and the fixed Stars emit light, the Heavens transmit it, and the Earth, the Planets, and the Comets reflect it: we shall, without being inaccurate, ascribe this threefold difference in visual appearance to three elements.

53.   That three heavens can also be distinguished in it.

And it will not be incorrect if, {from now on}, we take all the matter in the space AEI, which rotates around the center S,[50] to be the first heaven; and all that which forms a very great number of other vortices about the centers F, f, {and others similar}, to be the second; and finally, all that is beyond these two heavens, to be the third. Moreover, we think that the third one is immense in comparison to the second, just as the second is extremely large in comparison to the first. I shall, however, have no occasion here to discuss this third heaven, because it can in no way be seen

[49] The term for 'heavens' is capitalized throughout this article, for no apparent reason.
[50] See Plate VI.

by us during this life, and because I have only undertaken to speak of the visible world. However, we take all the vortices which there are around the centers F, f, to be only one heaven, {because they do not appear different to us, and} because they are considered by us in only one way. But as for the vortex the center of which is marked S, although it is not shown here to be an individual heaven different from the others, we nevertheless take it to be a separate heaven, and indeed the first or principal one: because we shall shortly discover the Earth, our dwelling, situated in it, and consequently, we shall have many more things to notice in it alone than in {both of} the others. For we are not accustomed to naming things for their own sake but in order to explain our thoughts about them; {and we ought generally to pay more attention to how they affect us than to what they in fact are}.

54.   How the Sun and the fixed Stars were formed.

Now, since the parts of the second element were wearing each other down with constant movement from the beginning, the matter of the first, {which must have formed from the scrapings from their angles}, gradually increased. And when there was more of it in the universe than was needed to fill the tiny spaces which are found between the adjacent spherical particles of the second element, whatever remained (after these spaces were filled) flowed toward centers S, F, f; and formed certain spherical and very fluid bodies there: namely, the Sun at center S, and fixed Stars at the other centers. For after the particles of the second element had been worn down more {and had become spherical}, they occupied less space than before. And thus they did not extend all the way to the centers, but, receding equally on all sides from them, left spherical spaces which were {immediately} filled by the matter of the first element flowing in from all the surrounding places;[51] for it is a law of nature that all bodies which are moved circularly attempt to recede from the centers around which they revolve.

55.   What light is.

I shall now try to explain, as accurately as I can, {the nature of} the force by which the {little} globules of the second element, and also the matter of the first which has accumulated around the centers S, f, {around which they

---

[51] The remainder of this sentence appears in the Latin text as the beginning of Article 55. The French has it here, which seems more appropriate.

revolve}, attempt to move away from these centers. For I intend to show further on that light consists entirely in this force {or effort}; and many other things depend on this knowledge.

56.   What ought to be understood about the striving of inanimate objects toward motion.

When I say that these little globules strive, {or have some inclination}, to recede from the centers around which they revolve, I do not intend that there be attributed to them any thought[52] from which this striving might derive; I mean only that they are so situated, and so disposed to move, that they will in fact recede if they are not restrained by any other cause.

57.   How there can be strivings toward diverse movements in the same body at the same time.

Now, inasmuch as it often happens that several different causes act simultaneously against the same body and some impede the effect of others; depending on whether we consider the former or the latter, we can say that this body strives or tends to move in {several} different directions at the same time. For example, the stone A,[53] when rotated in the sling EA around E, definitely tends from A toward B,[54] if all the combined causes which determine its movement from A to B are considered simultaneously; because it is in fact thus transported. But we can also say that, in accordance with the law of motion explained previously, the same stone tends toward C when it is at point A, if we consider only the force of its own movement {and agitation}; assuming AC to be a straight line which is tangent to the circle at point A. For {it is certain that}, if this stone (coming from L) were to emerge from the sling when it reached point A, it would go toward C and not toward B; and the sling, though it impedes this effect, does not impede the striving {toward C}. Finally, if instead of considering all the force of its motion, we pay attention to only one part of it, the effect of which is hindered by the sling; and if we distinguish this from the other part of the force, which achieves its effect; we shall say that the stone, when

---

[52] Latin: 'cogitatio'.
[53] See Plate VIII, Fig. i.

at point A, strives[54] only [to move] toward D, or that it only attempts to recede from the center E along the straight line EAD.[55]

58.   How bodies which are rotated strive to move away from the center of their movement.

In order that this may be clearly understood, let us compare the motion by which this stone when at A would go toward C (if it were impeded by no other force) with the motion by which an ant, starting at the same point A, would also be moved toward C.[56] We are assuming EY to be a rod on which the ant would be walking in a straight line from A toward Y, while the rod was being rotated around the center E, and its point A would thus describe the circle ABF. We are also assuming the rotation of the rod to be so proportioned to the motion of the ant that the ant would be at point X when the rod was at C, then at point Y when the rod was at G, and so on. Thus, the ant itself would always be on the straight line ACG.[57] Let us

[54] This does not seem to be in accord with the definition of 'strives' implicit in Article 56, since the stone is being restrained. Further, Descartes insists that, if unrestrained, the stone will move from A to C; thus, that is the path along which it "strives".

[55] The argument here is both complex and faulty. Descartes seems to wish to regard the stone's tendency to move from A to C as being composed of two other tendencies; one to move from A toward B, which the sling does not affect; and another to move away from E, which the sling prevents. Thus, the effect of the sling is to prevent "motion" from B to C. Article 58 contains an example for which such an analysis would be roughly correct. However, while it is true that *if* the stone tended to move toward B and away from E, it would move toward C (if the forces were of appropriate magnitude); it does not follow that it has either tendency. In fact, the stone has an inertial tendency from A toward C and a centripetal tendency toward E, produced by the pull of the hand; and, thus, the sling does affect the tendency toward C. But the stone does not tend to move toward C *when it is at B*; a fact upon which Descartes himself insists in Part II, Article 39. Indeed, Descartes seems to realize that there are no circumstances under which the stone will move directly away from E; but a tendency that is never realized is a strange sort of tendency. Descartes's analysis here is strikingly similar to one which he uses to demonstrate the law of sines for refracting media in the *Dioptrics*. There he considers the unrestrained path of a light ray as having a horizontal and a vertical component and derives the path which results from a hindrance to one component by the refracting medium. Here he considers the resultant motion from A to B as having an unrestrained component (toward C) and a restrained vertical component (away from E). See *Le Monde*, Chapter XIII, for a clearer version of this argument.

[56] See Plate VIII, Fig. ii.

[57] The inappropriateness of the comparison is clear. In this case, the ant *is* striving toward the end of the rod. But that would be so whether the rod was rotated or not; thus, rotation has nothing to do with the ant's effort away from E. The effect of the rotation is simply to make the ant's effort *describable* as "directed away from the center of rotation".

compare the force by which the same stone, being rotated in the sling and describing the circle ABF, strives to recede from the center E along the lines AB, BC, and FG, with the effort which the same ant would make if it were attached by some bond to the rod EY at point A. Then while the rod transported it around the center E, the ant would use all its strength in an attempt to move toward Y, and thus to recede from E along the straight lines EAY, EBY, and others similar.

59.   How great the force of this striving is.

I know the motion of this ant must indeed be very slow at the beginning, and its striving, judged only by this initial motion, cannot seem very great; yet the striving cannot be said to be nonexistent, and, since it increases as it produces its effect, the movement which results can become quite rapid.[58] But, {in order to avoid any possible difficulty}, let us use yet another example. {Let us see what will happen} if the little globe A is placed in the tube EY;[59] when we first begin to rotate this tube around the center E, the globe will advance only slowly toward Y. But in the next instant it will advance a bit faster, because in addition to retaining its original force it will acquire new force from its new striving to recede from E: because this striving continues as long as the circular motion lasts and is, as it were, renewed constantly.[60] Experience confirms this, for we see that when we rotate this tube EY very rapidly around the center E, the little globe which is inside it moves very promptly from A to Y.[61] We see, too, that the stone which is in a sling makes the rope more taut as the speed at which it is rotated increases; and, since what makes the rope taut is nothing other than the force by which the stone strives to recede from the center of its movement, we can judge the quantity of this force by the tension.

[58] That is, assuming that the ant always retains the speed already acquired, further effort will produce a continuous increase in speed.
[59] See Plate VIII, Fig. iii.
[60] The actual path of the globe will be along the tangent; it moves toward Y *only* because Y approaches the tangent as the tube is rotated. When Y intersects the tangent, the globe will be pulled away from the tangent by the closed end of the tube. The globe's speed relative to the tube does increase as Y is approached; but this is so because the path along the tube comes to more nearly coincide with the tangential path as the tube is rotated.
[61] This is, of course, because Y intersects the tangent more quickly.

60. That all the matter of the heavens strives similarly {to move away from certain centers}.

What I have just said about the stone rotating in a sling around the center E, or about the little globe in the tube EY, can easily be similarly understood of all the globules of the second element: i.e., each of these strives to recede from the center of the vortex in which it rotates; for it is restrained by the other globules beyond it in the same way as the stone is {restrained} by the sling.[62] Moreover, {it should be noticed that} the force of these globules is greatly increased by the fact that those higher up are pushed by those lower down, and all are simultaneously pushed by the matter of the first element accumulated in the center of this vortex. However, in order that all things may be clearly distinguished, I shall examine {the effect of} these globules separately, without considering the matter of the first element any more than if all the spaces which that matter occupies were empty; that is, if they were filled by a material which neither contributed anything to the motion of other bodies nor in any way impeded it. For in accordance with what has already been said, there obviously can be no other correct idea of empty space.

61. That this striving causes the bodies of the Sun and the fixed Stars to be spherical.

{First of all}, from the fact that all the globules which rotate around S in the vortex AEI[63] attempt to move away from the center S, as has already been shown, we can conclude that those situated on the straight line SA all push one another toward A, and that those situated on the straight line SE push one another toward E, and so on; so that if there are not enough of them to occupy all the space between S and the circumference AEI, all the unoccupied space will be in the vicinity of S. And inasmuch as those which rest upon one another (for example, those which are situated on the straight line SE) do not rotate in a body, like a rod, but complete their revolutions in varying lengths of time (as I shall explain later);[64] the space which they leave around S must be spherical. For even if we were to imagine that the

[62] Descartes is on somewhat sounder ground here. A rotating fluid does exert centrifugal force against the sides of its container, but once again, this is not the result of a tendency of the fluid to move directly away from the center of rotation but is the result of the container continually deflecting the fluid from the tangential path.
[63] See Plate VIIII, Fig. i.
[64] See Articles 83 and 84.

line SE in the beginning contained more globules than the lines SA or SI (so that the globules at the end of the line SE[65] would be nearer to the center S than those at the end of the line SI): nevertheless, the globules closest to S would complete their revolution before the more distant ones; and so some of them would go and place themselves at the end of the line SI, in order to move farther away from the center S. That is why {we must conclude that} they are now arranged in such a way that all those at the ends of the lines are equidistant from point S, and {that} consequently the space BCD, which they leave around this center, is spherical.

62.   That the same thing causes the heavenly matter to strive to recede from all the points on the circumference of each star and of the Sun.

In addition, it must be noticed that not only do all the globules situated on the straight line SE push one another toward E, but also each is pushed by all the others situated between the straight lines drawn from one of these globules tangent to the circumference BCD. Thus, for example, the globule F is pushed by all those situated between the lines BF and DF, or, in other words, inside the triangle BFD, but is not pushed by any of those outside this triangle. Therefore, if the space marked F were empty, all those which are in the space BFD, and no others, would simultaneously advance as much as possible in order to fill it. And, moreover, just as we see that the weight which carries a rock in a straight line toward the center of the earth, when it is in the air, makes it roll sideways when it falls down a slope; so, similarly, we must think that the force which makes the little globes in the space BFD tend to move away from the center S along straight lines drawn from that center can also make them move away from the center S along lines deviating {somewhat} from that center.[66]

63.   That the little globes of the heavenly matter do not impede one another in this striving.

And this example of weight will make my point very clearly understood if we imagine some balls of lead arranged like those shown in the flask

[65] That is, those at C.
[66] This modification of the view that rotating bodies tend directly away from the center is necessary to explain why light reaches us from all points on the sun's surface; see Article 64.

BFD,[67] which press one another in such a way that, when an opening is made in the base F of the vessel, globule 1 will descend {as much} by the force of its own weight {as by that of the others above it}. And at the moment when globule 1 begins to move {we shall see that} the two others 2, 2, will follow it, and the three others 3, 3, 3, will follow them, and so on. So that {we will also be able to see that} all those globules contained within the triangle BFD will immediately descend, but that none of those balls situated outside that triangle will move {in that direction}. It is true that, {in this example}, the two balls 2, 2, {come into contact} after dropping somewhat, {and thus} prevent each other from descending further. But the same thing is not true of the little globes which form the second element: for, although it sometimes happens that they are arranged like those shown in this figure, they however only remain in this position for that little space of time we call an instant, because they are perpetually in motion and so continue to move without interruption. Moreover, it must be noticed that the force of light does not consist in any duration of motion, but only in the pressing or first preparation for motion, even though actual motion may not result from this pressure.

64.   That all the properties of light are found in this striving: so that, by means of it, light, apparently coming from the stars, would be visible even if there were no force in the stars themselves.

From these things, it is clearly perceived how that action which I take to be light is emitted equally in all directions from the body of the Sun or of any fixed Star, and how it is transmitted in the shortest space of time to the greatest distance: and how this [occurs] along straight lines, drawn not only from the center of the luminous body, but also from any other point on its surface.[68] From this, all the other properties of light can be deduced. And, though this may perhaps seem a paradox to many people, all these things would exist in the heavenly matter, even if there were absolutely no force in the Sun or other star around which it was rotating. So that, if the body of the Sun were nothing other than an empty space, nevertheless, its light (which would admittedly not be as strong, but which would otherwise not differ from what it now is) would still be perceived by us. At least, {this

[67] See Plate VIIII, Figs. ii and iii.
[68] Thus, on Descartes's view, light is simply the pressure, or tendency to centrifugal motion, of the parts of transparent bodies; it does not involve the motion of anything physical from the sun or stars to the eye.

would be true} in the circle along which the matter of the heaven is rotated; for we are not yet considering here the other dimensions of the sphere {which extend toward the poles}. However, in order that we may be able to explain what there is in the Sun itself, and in the Stars, by which this force of light is increased and transmitted through all dimensions of the sphere, we must first say some things concerning the movement of the heavens.

65.   That the poles of each of the vortices of the heavens touch those parts of other vortices which are distant from their poles.

No matter how these individual vortices were moved in the beginning, they must now be arranged in harmony with one another so that each one is carried along in the direction in which the movements of all the remaining surrounding ones least oppose it. For the laws of nature are such that the movement of each body is easily turned aside by encounter with another body. Accordingly, if we suppose that the first vortex,[69] the center of which is S, is rotated from A through E toward I, the other vortex near to it, the center of which is F, must be rotated from A through E toward V if no other nearby vortices prevent this; for thus are their movements most compatible. And in the same way, the third vortex, which has its center, not on the plane SAFE, but above it (forming a triangle with the centers S and F), and which is joined to the other two vortices AEI and AEV on the line AE, must be rotated from A through E upward. This being so, the fourth vortex, the center of which is f, cannot be rotated from E toward I, to make its movement compatible with that of the first, because it would thereby oppose the movements of the second and third; nor [can it be rotated] from E toward V, like the second, because the first and third would oppose it; nor, finally, [can it be rotated] upward from E, like the third, because the first and second would oppose it. Therefore, the only remaining possibility is for it to have one of its poles at E and the other on the opposite side at B, and to be rotated on its axis EB from I toward V.

66.   That the movements of these vortices are deflected in some way, so that they may be in harmony.

In addition, we must notice that there will still be some contrariety in these movements if the ecliptics, that is to say, the circles furthest from the

[69] See Plate VI.

poles,[70] of the first three vortices meet directly at point E, {where I place} the pole of the fourth.[71] If, for example, IVX is that part of the fourth which is near the pole E and is revolved according to the order of the markers IVX; the first vortex, rubbing against this part along the straight line EI and others parallel to it, and the second vortex, rubbing against it along the straight line EV, and the third, rubbing against it along the line EX, would all oppose its circular movement somewhat. But nature corrects this very easily through the laws of motion, by deflecting somewhat the equators of these three vortices in the direction in which the fourth one, IVX, is rotating; so that these three afterwards rub against it, not along the straight lines EI, EV, and EX, but along the curved lines 1I, 2V, and 3X, and thus adapt themselves completely to its movement.

67.   That two vortices cannot touch at their poles.

I do not think that any better way to reconcile the movements of several vortices can be found. Because, if we suppose that there are two which touch at the poles, either they will both turn in the same direction and thus will unite to form a single vortex, or else they will turn in opposite directions and thus oppose each other as much as possible. That is why, although I am not daring to presume to determine the situations and movements of all the vortices of the heaven; I nevertheless think that I can affirm, in general, that each vortex has its poles further from the poles of those nearest to it than from their equators. And I think that I have demonstrated this sufficiently here.

68.   That these vortices are unequal in size.

Further, that inexplicable variety which appears in the situation of the fixed Stars seems to show clearly that those vortices which are rotated around them are not equal to one another [in size]. However, I judge that it is obvious, from their light, that no fixed Star can exist anywhere except in the center of some such vortex: for this light can be explained very exactly by means of such vortices, but in no other way without them; as will be

---

[70] From this point on, Descartes frequently refers to the "ecliptic" of a rotating sphere when he means its equator. Henceforth, 'ecliptic' will be replaced by 'equator' when that is clearly the intended meaning. Throughout Part III, Descartes seems to assume that the sun's equator, the equator of the sun's vortex, and the ecliptic all lie in roughly the same plane; this is never explicitly asserted, however.

[71] See Plate VIIII, Figs. iv and v.

obvious partly from what has already been said and partly from what is to be said later. And since, by means of our senses, we perceive absolutely nothing in the fixed Stars except their light and apparent situation, we have no reason to attribute to them anything other than what we judge necessary to explain these two things. And nothing more is required to explain their light than for the vortices of heavenly matter to rotate around them; similarly, nothing more is needed to explain their apparent situations than for their vortices to be unequal in size. But if they truly are unequal, it is necessary for those parts which are remote from the poles of some vortices to touch other vortices at points which are near the poles [of the others]: because the similar parts of larger and smaller vortices cannot {all} be contiguous to one another.

69.   That the matter of the first element flows from the poles of each vortex toward the center, and from the center toward the other parts.

From these things, it can be known that the matter of the first element flows continuously from the other surrounding vortices toward the center of each vortex through the parts adjacent to its poles; and that, *vice versa*, this matter flows out of each vortex, into the other surrounding ones, through the parts remote from its poles. Let us suppose that AYBM is the vortex of the first heaven,[72] in the center of which is the Sun, and that B and A are, respectively, its North and South poles, around which the whole [vortex] rotates; and that the four surrounding vortices K, O, L, and C rotate around their axes TT, YY, ZZ, and MM, in such a way the AYBM touches the two {designated as} O and C at their poles, and the other two, K and L, at parts very distant from their poles. It is obvious from what has been said before that all the matter of AYBM strives to recede from the axis AB, and therefore tends with greater force in the directions of Y and M than in the directions of A and B. And since at Y and M it encounters the poles of the vortices O and C, where there is no great force to resist it; while at A and B it encounters those parts of the vortices K and L which are most remote from their poles (and consequently have more force to travel from L and K toward S than the parts surrounding the poles of the vortex S have to travel toward L and K): there is no doubt that the matter which is at K and L must proceed toward S, and that which is at S, toward O and C.

[72] Articles 69 through 80 refer to Plate X.

70.   That the same cannot be understood of the matter of the second element.

Indeed, this would also have to be understood, not only of the matter of the first element, but of the globules of the second, if no individual causes impeded their movement in that direction. But because the agitation of the first element is much more rapid than that of the second, and because there is always free passage for it through those narrow angles which cannot be occupied by the globules of the second; even if we were to imagine that all the matter, of both the first and second elements, contained in the vortex L had begun to proceed simultaneously toward S from a place midway between the centers S and L, we would however understand that the matter of the first element would have had to reach the center S more rapidly than that of the second. Then, having thus entered the space S, the matter of the first element drives the globules of the second with so much force, not only toward the equator eg or MY, but primarily toward the poles fd or AB, as I shall soon explain, that it thereby prevents those globules coming from vortex L from getting any closer to S than a certain boundary marked here by the letter B. And the same is to be concluded about the vortex K and all the others.

71.   The reason for this diversity.

In addition, it is also important to consider [the fact] that the particles of the second element which are rotated around the center L not only have the force to recede from this center, but also to continue to move at the same speed. There is a certain opposition between these two things; because while these particles are rotating in the vortex L, they are confined within certain limits by other neighboring vortices which must be understood to be above and below the plane of this figure. Therefore, they cannot advance toward B {where their space is not as limited} unless they move more slowly {when} between L and B than {when} between L and the other neighboring vortices understood to be outside the plane of this figure; and indeed, more slowly in proportion as the space LB is larger.[73] For, since they are moved circularly, they cannot take more time to pass between L and the other vortices, than between L and B. And consequently, the force which they have to recede from the center L causes some of them to advance toward B,

[73] See Part II, Article 33.

because there they encounter the parts around the poles of the vortex S, which yield to them without difficulty. On the other hand, the [lesser] force which they have to retain the speed of their movement prevents them from advancing far enough to reach S. The same thing is not true of the first element: for, even though it resembles the particles of the second in that, while rotating simultaneously with them, it strives to move away from the centers of the vortices in which it is contained; it is, however, very different in that it does not need to lose any of its speed when it recedes from these centers, because it everywhere finds passages {of} almost equal {size} which permit it to continue its movements. These passages are in the narrow angles which are not occupied by the globules of the second element. Therefore, there is no doubt that this matter of the first element continuously flows toward S through the parts near the poles A and B; not only from the vortices K and L, but also from many others which are not shown in this figure; because they are not all to be understood to be on the same plane, and because I cannot determine either their situation, their magnitude, or their number. There is no doubt either, that the same matter flows out of S toward the vortices O and C, and also toward others, whose situation, size, and number I do not establish. Nor do I establish whether this same matter is immediately returned back from O and C to K and L, or whether it is rather diverted to many other vortices, more remote from the first heaven, before completing the circle of its movement.

72.   How the matter which forms the Sun is moved.

But we must consider a little more carefully how this [matter] is moved in the space defg. Of course that part of it which has come from A proceeds in a straight line to d, where it encounters the globules of the second element and drives them toward B. And in the same way the other part, which has come from B, proceeds in a straight line to f, where it encounters the globules of the second element and drives them back toward A. And immediately, both the matter which is at d, and also that which is at f, is everywhere turned back toward the equator eg, and drives all the surrounding globules of the second element equally; and, finally, it escapes toward M and Y through the passages which are between those globules around the equator eg. Furthermore, while this matter of the first element is being carried along in a straight line by its own motion from A and B toward d and f, it is also being carried along circularly by the movement of the entire vortex around the axis AB; so that these individual scrapings describe lines which are spiral, or twisted like a cochlea. And after these spirals have reached d and f, they are turned back from there, on both sides, toward the equator eg. And, because the space defg is larger than the passages through which the matter of the first element enters and leaves it, some part of that matter consequently always remains there and forms a very fluid body which is perpetually rotated around the axis fd.

73.   That there are various inequalities in the situation of the body of the Sun.

And we must particularly notice that this body must be spherical. Because of the inequality of the vortices, it must not be thought that the quantity of the matter of the first element which is sent toward S by the vortices near one of its poles is exactly equal to that sent by those near the other, nor that these vortices are so located that they send this matter in exactly opposite directions. Nor should we think that those other vortices which touch the first heaven near its equator {have their poles so arranged that they} all face that circle of the equator in the same way or that these vortices admit the matter leaving from S equally through all the locations on and near to that circle. However, from this there can be proved no inequalities in the shape of the Sun, but only in its situation, movement, and quantity. Specifically, if the force of the matter of the first element coming from pole A toward S is greater than that of the matter coming from pole B; before the former matter can be driven back by encounter with the other, it will proceed further toward B than the other toward A. But by thus proceeding further, its force will be decreased; and, according to the laws of nature, both will finally drive each other back at that place where their forces will be perfectly equal, and there they will form the body of the Sun: which consequently will be further from pole A than from pole B. But the globules of the second element are not driven with greater force at point d on its circumference than at point f, and therefore this circumference will be no less round. Similarly, if the matter of the first element leaves S more easily in the direction of O than in the direction of C (because the former space is less restricted), that alone will make the body S approach O somewhat, and thereby somewhat diminish the intervening space, and at last come to rest where the force is equal on both sides. And thus, even though we were considering only the four vortices L, C, K, and O; provided only that we suppose them to be unequal in size, it follows from this that the Sun S must be neither halfway between O and C, nor halfway between L

and K. And still greater inequality in its situation can be understood from
the fact that many other vortices are situated around it.[74]

74.   That there are also various inequalities in the motion of its
                       matter.[75]

Further, if the matter of the first element coming from the vortices K and
L is not as inclined to move directly toward S as toward other places near
there; for example, if that matter which comes from K is more inclined to
move toward e,[76] and that which comes from L to move toward g; that will
prevent the poles f, d, around which all the matter of the Sun revolves, from
being on the straight lines drawn from K and L to S. Instead, the South pole
f will move somewhat toward e and the North pole d toward g. In the same
way, if the straight line SM, along which {I am supposing} the matter of the
first element moves more easily from S toward C {than along any other},
passes through a point on the circumference fed which is closer to d than to
f; and if, similarly, the line SY, along which {I am supposing that} this
matter travels from S to O, passes through a point on the circumference fgd
which is closer to f than to d: that will cause gSe, which here represents the
equator of the Sun, that is to say, the plane on which moves that part of its
matter which describes the greatest circle, to be more inclined at e toward
the pole d than toward the pole f, but still by no means as much inclined as
the straight line SM is. And it will be more inclined at g toward f than
toward d, but, again, by no means as steeply as the straight line SY. From
which it follows that the axis around which revolves all the matter which
forms the body of the Sun and which is defined by the poles f and d is not
exactly straight but a little curved {on both sides}; and that this matter turns
a little more quickly between e and d or between f and g, than between e and
f or d and g; and that perhaps, too, the speed at which it revolves between e
and d is not precisely equal to that at which it turns between f and g.[77]

[74] This may be an attempt to account for the fact that, in the Copernican system, the sun is not
located exactly at the center of the planetary orbits.
[75] The French text has been preferred here; both because it is clearer, and because it is in
accord with the explanation of this article which Descartes gave Picot in February, 1645. See
A. & T., IV, 181–182.
[76] This occurs because the plane of the equator of K's vortex intersects the sun at e rather than
at its center, and thus the matter from K tends to enter the sun on that side, thereby moving the
sun's poles toward e.
[77] Thus, the location of the poles and equator of the sun depends partly on the relative location
of the poles and equators of the neighboring vortices as well as on the location of the poles and
equator of the sun's own vortex. All this may be an attempt to account for the fact that
sunspots appear to move in curved lines which are inclined to the ecliptic.

75.   That these inequalities do not however prevent its shape from
             being spherical.

This cannot however prevent its body from being as nearly spherical as possible; because, meanwhile, its other movement from the poles toward the equator compensates for these inequalities. And [this happens] in the same way as we see a glass bottle become spherical solely from the fact that air is blown through an iron tube into its molten matter: because of course this air does not move from the mouth of the bottle toward its base with more force than [that with which] it is deflected from there to all other parts, and drives them all equally easily. Thus, the matter of the first element, having entered the body of the Sun through its poles, must drive back all the surrounding globules of the second element equally on all sides; those toward which it is only obliquely directed no less than those it strikes directly.

76.   Concerning the movement of the first element while it is
             situated among the globules of the second.

It must next be noted that this matter of the first element, while situated among the globules of the second, certainly has a movement, along straight lines, from the poles AB to the Sun and from the Sun to the equator YM, and {shares in} a circular movement around the poles common to the whole heaven AMBY. But it expends the maximum and principal part of its own agitation in constantly changing the shapes of its tiny parts, so that they may exactly fill all the narrow spaces through which they pass. This causes its force, which is very much divided, to become weaker; and its individual particles to accommodate themselves to the movements of the globules of the second element near to them; and to be constantly prepared to leave these narrow spaces in which they are forced to move so obliquely, in order to proceed in a straight line in any direction. However, that matter which has been collected in a mass in the body of the Sun has very great force there, because of the uniting of all its parts in the same very rapid movements, and it expends all this force in driving the surrounding globules of the second element this way and that.[78]

---

[78] That is, the matter of the first element has sufficient force to drive the particles of the second primarily when it is concentrated in a sun. But cf. Article 84.

77.   How the light of the Sun extends not only toward the Ecliptic, but also toward the poles.[79]

And from these things it can be understood how much matter of the first element unites for that action in which we previously showed light consists, and how that action extends in all directions; not only toward the ecliptic, but also toward the poles. For first, if we think that there is at H a space which is entirely filled by the matter of the first element,[80] and yet large enough to admit one or more of the globules of the second element; there is no doubt that all the globules contained in the cone dHf, the base of which is the concave hemisphere def, will simultaneously move toward H.

78.   How it extends toward the Ecliptic.

Now this [tendency toward H] has been shown above,[81] regarding the globules contained in the triangle whose base was the semi-circle of the solar equator, although no action of the first element was yet being considered; but now this same thing will, by considering [the action of] the first element, be very clearly revealed, not only of those globules, but, at the same time, also of the remaining ones contained in the whole cone.[82] For that part [of the matter of the first element] which forms the body of the Sun not only drives the globules of the second element which are at the equator e toward H, but also those which are at the poles d and f, and finally all those which are in the cone dHf; for it is not moved with greater force toward e than toward d, and f, and the other areas between. Indeed, the matter which we are now supposing to be at H moves toward C, and from there, via K and L, to S, returning as it were, in a circle. Therefore, it does not prevent these globules [of the second element] from reaching H; and by their approach, the space where they were before is added to the body of the Sun, and filled by the matter of the first element flowing in from the centers K, L, and others similar.

[79] Articles 77 through 81 represent Descartes's attempt to overcome an unfortunate apparent consequence of his view of the nature of light. If light results from the centrifugal force of the particles of the sun, it should be emitted only at right angles to the axis of rotation. Indeed, since the particles move more slowly the nearer they are to the poles, the sun should be brightest at its equator and dimmer toward the poles.
[80] H, it should be noted, is near the sun's equator.
[81] See Article 62.
[82] That is, the cone with H as its vertex and the hemisphere of the sun as its base.

79.   How, in response to the movement of one very small body, others extremely distant from it are easily moved.

Indeed, {far from preventing the parts of the second element from advancing to H}, this matter [of the first element] rather contributes to the effect. For, seeing that all movement tends along straight lines, the most agitated matter which is at H has more propensity to leave H than to remain there; for the narrower the space in which it is situated, the more it is forced to curve its movements. And therefore we ought not to be at all astonished that, often, in response to the movement of some extremely tiny body, other bodies, scattered over as great a space as you like, are simultaneously moved: and accordingly, we ought not to wonder, either, why the action, not only of the Sun, but also of most distant Stars, reaches the earth in the shortest space of time.

80.   How the light of the Sun moves toward the poles.

{Having thus seen how the Sun acts in relation to the Ecliptic, we can now see in the same way how it acts in relation to the poles}, if we suppose there to be some space at point N which is filled solely by the first element, {although it is large enough to contain some of the parts of the second}. For, since the matter which forms the body of the Sun is moved from d toward f and simultaneously toward the entire hemisphere efg, we shall easily understand that it must drive all the parts of the second element contained in the cone gNe toward space N: because these parts, although they may in themselves have no inclination to move in that direction, are not disposed to resist the action which drives them. The matter of the first element which fills space N does not prevent them {from entering that space}, for it is entirely inclined to {leave it and} move toward S to fill the space which they leave on the Sun's surface efg, as they move toward N. Nor is there any difficulty in the fact that the matter of the second element {which is in the cone eNg} advances {in a straight line} from S to N while that of the first element simultaneously moves in the opposite direction, from N to S. For, since this matter of the first element passes only through those very narrow interstices which the globules of the second do not fill, its movement cannot {hinder theirs or} be hindered by them. In a similar fashion, we see that in hour-glasses (which we now use instead of water-clocks) the air contained in the lower vessel is not prevented from rising into the upper one through the interstices between the grains of sand falling down.

81.   Whether its force is equal at the poles and at the ecliptic.

It remains only to ask whether the globules contained in the cone eNg are driven toward N, solely by the matter of the Sun, with as much force as those in the cone dHf are driven toward H by the same matter of the Sun plus their own movement, {which tends to move them away from the center S}. It does indeed seem {very likely} that this force is not equal, if we suppose H and N to be equidistant from {point} S. But since I have already observed that the distance between the Sun and the circumference of the heaven which surrounds it is less toward the poles than toward the ecliptic: {we should judge, I think, that} that force can be equal at the highest point [of the cones] when the ratio between lines HS and NS is equal to that between MS and AS. And we have only one phenomenon in nature from which we may obtain an experience of this matter. If a Comet passes through such a great part of the heaven that it is first seen at the ecliptic, then near one of the poles, and subsequently at the ecliptic again; then, having determined the ratio of its distances, we can judge whether its light (which, as I shall show later, comes from the Sun), *caeteris paribus*, appears stronger at the ecliptic than at the poles.

82.   That up to a certain distance, the globules of the second element
             which are close to the Sun are smaller and move more rapidly
             than those which are further away; beyond which distance, all
             are equal in magnitude and move the more rapidly the further
             they are from the Sun.

It still remains to be noticed that the globules of the second element which are closest to the center of each vortex are smaller and move more rapidly than those which are a little further from it, and that this is the case only up to a particular limit, beyond which, those which are higher move more rapidly than those which are lower and are all equal in size. For example, it must be thought that here in the first heaven the tiniest globules of the second element are those which touch the surface of the Sun defg, and that those which are further away from it are progressively larger, {according to the different levels on which they are situated}, up to the irregular sphere HNQR; while those which are beyond this sphere are all equal in size. And those which move the most slowly of all are those in the area HNQR: so that the parts of the second element {which are at} HQ may take thirty years or more to describe a circle around the poles AB, while, on the other hand, those which are higher, toward M and Y, and those which are lower, toward e and g, move more quickly and take only a few weeks to complete their revolution.[83]

83.   Why those furthest away move more rapidly than those
             somewhat less distant.

To begin with, it is easy to demonstrate that those which are higher up, toward M and Y, must move more rapidly than those which are lower, toward H and Q. For from my supposition that they were all equal in the beginning {of the world} (and I think I was right to suppose this, having no proof that they were unequal), and from the fact that the space in which they are carried circularly, as in a vortex, is not exactly spherical (partly because the other surrounding vortices are not all exactly equal in size and partly because it must be narrower [in the areas] opposite the centers[84] of each of these [other] vortices than elsewhere); it follows that some of these parts must move more quickly than others when they must change their order so as to pass from a wider path to a narrower one. {As we can see here,}[85] the two balls which are between points A and B cannot pass between the other two points C and D, {which I am supposing to be closer together}, unless one moves ahead of the other; and it is obvious that [in order for this to occur], the one which precedes must move more rapidly than the other. Now, since all the globules of {the second element which form} the first heaven strive to move away from the center S;[86] as soon as there is one which moves more quickly than those around it, this speed gives it more force and causes it to go beyond them; so that those parts furthest from the center are always those which move the most rapidly. {I do not specify} their rate of speed, {because} we can learn {this} only by experience. Our only sources of such experience are Comets, which, as I shall show later, pass from one heaven into another, {and more or less follow the course of the one in which they are located}. Neither do I specify how slow is the movement of the circle HQ; for all we know about it is what

---

[83] This arrangement is necessary to explain why planets nearer the sun move more rapidly than those further away.

[84] See A. & T., V, 172, for Burman's account of Descartes's explanation of this. See also Plate VI, noting that the vortex H is more compressed between S and F than between S and E, say.

[85] See Plate XI, Fig. i.

[86] Articles 83 through 86 refer to Plate X.

we can learn from the course of Saturn, {which takes thirty years to complete}; I shall show later that Saturn is on or beneath that circle.

84.   Why those parts closest to the Sun are more rapidly moved than those a little further from it.

It is also easy to prove that among the parts of the second element contained in the area HQ, those closest to the center S must complete their revolution in less time than those which are further from it; because the Sun's rotation around the same center must carry them with it. For, inasmuch as it moves more rapidly than they do, and because some parts of its matter are constantly flowing out of it at the equator between those of the second element, while others are flowing in at the poles; it obviously must have the force to carry with it all these globules {of the surrounding Heaven}, up to a certain distance.[87] And the limits of that distance are here represented by the ellipse HNQR, rather than by a circle.[88] For, although the Sun is spherical and, by the action which I have said we must take to be its light, drives the parts of the heaven which are near its poles no less vigorously than those near its equator, yet the same is not true of that other action by which it circularly carries along those closest to it; because this depends solely on its circular movement around its axis and this undoubtedly has less force near the poles than near the equator. That is why H and Q must be further from the center S than are N and R; and this will later explain why the tails of Comets appear {to us} sometimes straight and sometimes curved.

85.   Why those nearest to the Sun are smaller than those further away.

Since below the boundary HQ, the parts of the second element which are very close to the Sun move more rapidly than those which are a little further away; they must also be smaller. For if they were larger or of equal size, that alone would give them more force and they would pass beyond the others, {since their speed is greater than those higher up}. But if it happens that one

---

[87] This would seem to claim that the matter of the first element can affect the matter of the second when it is outside the sun, in contrast to what seems to be claimed in Article 76.
[88] The ellipse HNQR does not represent a planetary orbit; those would be roughly at right angles to the page. Rather, it is a cross-section, taken through the poles of the sun, of the ellipsoid of matter which is affected by the sun's rotation.

of these parts is so small in proportion to those beyond it, that its additional speed, {resulting from its greater proximity to the Sun}, is less than {and does not increase its force by as much as} the additional size of these others {increases theirs; it is obvious that} it must always remain below them, {close to the Sun, even though it moves more rapidly}.[89] And, although I have supposed that God created all these parts of the second element equal in the beginning, some must have become smaller than others with the passing of time. This happened because of the inequality of the spaces through which they had to pass and the resultant inequality in their movement, as I have just shown; {and because this must thereby have resulted in some inequality in their size, since those which travelled at the greatest speed collided with each other more violently and thus lost more of their matter}. And those parts which became {noticeably} smaller than the others must be sufficiently numerous to fill the space HNQR; for we consider that space to be extremely small compared with the size of the whole vortex AYBM, as the size of the Sun must also be considered to be extremely small compared with AYBM {or with HNQR}. But these proportions could not be shown here because the figure would have had to be too large. And it should be noted that there are various other inequalities in the motions of the heavenly globules especially of those between S and H or Q, but it will be more appropriate to speak of these a bit later.

86.   That the globules of the second element are simultaneously moved in various ways, as a result of which they become perfectly spherical.

Finally, it must not be omitted that the matter of the first element, coming from vortices K, L, and others similar, is carried principally toward the Sun; however, very many of its parts are dispersed throughout the whole vortex AYBM, and from there cross to the other vortices C, O, and similar ones. And by flowing around the globules of the second element, they cause these to be moved both around their own centers and also perhaps in other ways. And since these globules are thus agitated, not merely in one way, but in many diverse ways simultaneously; we clearly perceive from this that, whatever shape they may have had in the

---

[89] That is, the centrifugal force of a particle is proportional to its quantity of motion, and those particles with the greater centrifugal force will move furthest from the center. However, cf. Article 121.

beginning, they must now be perfectly spherical, and not like a cylinder or like any spheroid which is round only from one aspect.

87.   That there are various degrees of agitation in the particles of the first element.

Now that we have thus in some way explained the nature of the first two elements; in order to be able, finally, to speak of the third, we must consider that the matter of the first is not equally agitated in all its particles, and that frequently, in a very small quantity of this matter, there are innumerable diverse degrees of speed. This can easily be shown, both from the way in which we earlier described its creation and from the function which it must continually fulfil. For we imagined that it was produced as a result of the fact that when the parts of the second element were not yet spherical but angular, and entirely filled the space which contained them; they could not have moved without breaking {the little points of} their angles, nor without whatever parts were detached from them, {as they became spherical}, variously changing their shape, in order to exactly fill the various spaces which were to be occupied: and thus they assumed the form of the first element. And its function is still to occupy in this way all the little spaces which are found between all other bodies. From this it is obvious that all these tiny particles must, in the beginning, have been no larger than the angles {which had to be separated from those of the second element to enable them to move}; or no larger than the space which three contiguous parts of the second element leave between themselves {now that they have become spherical}. Therefore, while some may have subsequently remained undivided, others must have been indefinitely divided into particles able to adapt themselves to the constantly changing sizes of the little spaces {which occur between the moving parts of the second element}. For example, if we assume that there are three globules A, B, and C,[90] and that the first two, A and B, which touch each other at point G, move only around their own centers, while the third, C, which touches the first at point E, rolls over {the surface of} the first one from E to I, until its point D comes in contact with point F of the second; it is obvious that the matter of the first element which is contained in the triangular space FGI, and which may consist either of several tiny particles or of only one, can meanwhile remain there motionless; but the matter which fills the space FIED must move.

[90] See Plate XI, Fig. ii.

Moreover, it is impossible to point out, between F and D, a single one of these scrapings, however small, which is not larger than the one which is removed {from the line FD} at each moment. Because, as globule C approaches B, it shortens line DF, causing it to be of innumerable different degrees of shortness successively.

88.　　　That those tiny particles which have the least speed easily transfer to others that which they have, and adhere to one another.

So therefore, in the matter of the first element, there are certain scrapings less divided than the rest and less rapidly agitated. And since we are supposing these scrapings to have been torn away from the angles of the particles of the second (when they had not yet been rounded into globules and they alone filled all spaces); it is impossible for these scrapings not to have extremely angular shapes, ill-adapted to movement. As a result, they easily adhere to one another and transfer a great part of their agitation to those other scrapings which are the tiniest and most rapidly agitated. Because, according to the laws of nature, other things being equal, larger bodies transfer their agitation to smaller ones more easily than they can receive any new agitation from these others. {Consequently, it can be stated that the smallest parts are usually the most agitated.}

89.　　　That such clusters of tiny particles are principally found in the matter of the first element which flows from the poles toward the centers of vortices.

Such parts {which thus become attached to one another and which retain the least agitation} are mainly found in that matter of the first element which is moved in straight lines from the poles of each heaven toward its center: for this movement in straight lines requires less agitation that the other more oblique and diverse movements which occur elsewhere. Thus, {when these parts [under discussion] are in these other places}, they are usually expelled into the path of this straight movement, where several of them unite to form certain little bodies the shape of which I wish to consider very carefully.

90.　　　What the shape of these particles, which from now on we shall call grooved, is.

Of course, they must be triangular in cross-section, because they frequently pass through those narrow triangular spaces which are created

when three globules of the second element touch. As for their length, it is
not easy to determine, since it seems to depend solely on the quantity of
matter of which these small bodies consist; but it suffices that we conceive
of them as small {fluted} cylinders with three grooves {or channels} which
are twisted like the shell of a snail. This enables them to pass in a twisting
motion through the little spaces which have the form of the curvilinear
triangle FGI[91] and which are always found between three contiguous
globes of the second element. For, since these {grooved particles} are
oblong and pass very rapidly between the particles of the second element
(while these are themselves being rotated around the axis of the heaven by
another movement), we can easily understand that the grooves of each one
must be twisted like the grooves in a snail's shell; and that these {three
grooves} are more or less twisted according to the distance which separates
the spaces through which they are passing from that axis; because the parts
of the second element revolve more rapidly when further from the axis than
when closer to it,[92] as was stated earlier.

91.   That the particles coming from opposite poles are twisted in
               opposite directions.

Moreover, because they approach the center of the heaven from opposite
directions, that is, some from the South {pole} and some from the North,
while the whole vortex rotates in one direction on its axis, it is obvious that
those coming from the South pole must be twisted in exactly the opposite
direction from those coming from the North. And this fact seems to me
very noteworthy, because the force of the magnet, which is to be explained
later,[93] mainly depends upon it.

92.   That there are only three grooves on the surface of each of these
               particles.

Moreover, it should not be thought that I have no reason for stating that
these parts of the first element have only three grooves on their surface. For

[91] See Plate XI, Fig. ii.
[92] Descartes seems a bit confused here. Close to the sun, the matter of the second element
rotates more rapidly near the axis, though this is not true beyond the orbit of Saturn. Thus, a
particle would seem to need to change its degree of twist when it enters the ellipsoid of matter
rotated by the sun.
[93] Part IV, Article 133, *et seq*.

though the globules of the second do not always touch one another in such a way as to leave triangular spaces between them, we can see here[94] that any other larger spaces between these parts {of the second element} always have angles precisely equal to those of the triangle FGI, and that, in addition, these spaces are constantly changing; so that the grooved particles which pass through them must assume the figure I have described. For example, the four globules A, B, C, H, touching one another at points K, L, G, E, leave between them a quadrangular space, each angle of which is exactly equal to each angle of the triangle FGI.[95] And because these four globules constantly change the figure of this space as they move, making it sometimes square and sometimes oblong, and sometimes causing it to be divided into two triangular spaces;[96] the less agitated matter of the first element situated in it must flow toward one or two of these angles and leave the remaining space to the more agitated matter which can change its shape {at any moment} in order to adapt to all the movements of these globules. Further, if there is by chance some particle {of this matter of the first element} which is situated in one of these angles but which extends toward the point opposite that angle beyond the space of the equilateral triangle FGI, it will be driven out of there and accordingly decreased in size whenever a third globule advances to touch the other two which form the angle in which this particle is situated. For example, if the less agitated matter which fills angle G extends toward D beyond the line FI, the globule C, as it approaches B, will drive it out of there and cut off that part which is preventing B from closing the triangle GFI. And because these parts of the first element, which are the largest and the least agitated, must very often, in their journeys about the heavens, find themselves between three globules which are thus advancing to touch each other; it seems that the only definite figure which they can retain for any length of time is the one which we have described.

93.   That between the grooved particles and the smallest particles of all, there are others in the first element which are of various sizes.

Now, although these grooved and oblong particles are very different from the remaining material of the first element, I nevertheless include

---

[94] See Plate XI, Fig. iii.

[95] That is, the space which would be left if C were touching both A and B.

[96] When B and C are in contact both with each other and with A and H, for example.

them all in {the category of} the first element while they are among the
globules of the second; both because we do not notice that they produce
any {different} individual effects there, and because we judge that, between
these grooved particles and the smallest ones, there are many others of
innumerable intermediate sizes and degrees of agitation, as can easily be
proved by the diversity of the places through which they pass {and which
they fill}.

94.   How the spots on the surface of the Sun or stars are formed
                 from these particles.

But when the matter of the first element reaches the body of the Sun or of
another star, all the most agitated of its tiny particles strive to unite there in
similar movements; because they are not hindered by any impediment of
the globules of the second element. As a result, these grooved particles (and
also many other slightly smaller ones, which resist so much agitation on
account of their overly angular shapes or excessive sizes) are separated
from the other tiny particles. Easily adhering to one another because of the
inequality of their shapes, they sometimes form very large masses, which,
being immediately contiguous to the surface of the heaven, are joined to the
star from which they emerged. There they resist that action in which we
stated earlier the force of light consists; and thus are similar to those spots
which are usually observed on the surface of the Sun. In the same way, we
see that water and other liquids, when they boil up after being placed near a
fire and contain some particles of a different nature from and less suited to
movement than the rest; give off a dense scum composed of these particles:
which generally floats on their surface and has very irregular and
changeable shapes. Thus it is evident that the matter of the Sun, bubbling
on both sides from its poles toward the equator, must reject like a scum the
grooved particles and all those others which easily adhere to one another
and have difficulty in complying with its common motion.

95.   That from this the principal properties of these spots are
                 learned.

And from this it is easy to know why the Sun's spots do not usually
appear near its poles, but rather in the areas near the equator; and why they
have extremely varied and indeterminate shapes; and finally why they are
moved in a circle around the axis of the Sun, if not as rapidly as its
substance, at least along with that part of the heaven closest to it.

96.   How these spots are destroyed and new ones formed.

However, in the same way as many liquids, by boiling longer, reabsorb and consume the same scum which they gave off in the beginning by bubbling up; it must be similarly thought that with the same facility with which the matter of the spots emerges from the body of the Sun and accumulates on its surface, it is shortly afterwards diminished; partly drawn back into the Sun's substance, and partly dispersed throughout the nearby heaven.[97] (For these spots are not formed from the whole body of the Sun, but only from the matter which has recently entered it.) And the other matter which has been longer in the Sun (and is now, so to speak, purified by heat and clarified), by constantly rotating with great speed, partly wears away those spots which have already been formed; while new ones are being created elsewhere from the new matter entering the Sun: as a result of which they do not all appear in the same places. And certainly the entire surface of the Sun, with the exception of the regions around its poles, is generally covered by the matter from which they are formed; but there are said to be spots only in those places where this matter is so dense and compact that the force of the light coming from the Sun is perceptibly weakened.

97.   Why the colors of the rainbow appear at the edges of some
           spots.

Furthermore, it can happen that these spots, when slightly thicker and denser, are worn away at their circumference sooner than in the center by the purer matter of the Sun flowing around them; and consequently that the extremities of their circumferences, growing thinner, allow the Sun's light to pass through. From this it follows that these [edges] must be tinted by the colors of the rainbow, as I previously explained concerning a glass prism in the eighth discourse of my *Meteorology*.[98] And such colors are sometimes observed in them.

98.   How the spots become *faculae*,[99] and *vice versa*.

It also often happens that the matter of the Sun which is flowing around these spots surges over their extremities, and, cut off between them and the

[97] Sunspots normally last less than a week.

[98] The reference is to the *Discourses on Meteorology*, published in 1637 as part of the *Discourse on Method*.

[99] *Faculae* are areas on the sun's surface which are brighter than normal. They usually develop into sunspots and re-appear after a sunspot has faded away.

surface of the nearby heaven, is then forced to move more rapidly than usual: in the same way as the rapidity of rivers is always greater in shallow and narrow places than in deep and wide ones. From this it follows that the Sun's light must be somewhat stronger there.[100] Thus the spots are often transformed into *faculae*; that is, certain parts of the Sun's surface which previously were darker {than the rest}, subsequently become brighter. Conversely, the *faculae* are seen to be changed into spots when the latter have been submerged on only one side in the more subtle matter of the Sun, and when a great quantity of new matter [suitable for forming spots] accumulates on the other side and adheres.

99.   The kind of particles into which these spots disintegrate.

However, when these spots are disintegrated, they are not transformed into particles exactly similar to those from which they were formed: but some are smaller and at the same time more solid, or less angular in shape. These, consequently, are better suited to movement, and therefore easily move toward other vortices through the passages between the globules of the surrounding heaven. Others are extremely small and have been worn off the angles of the former. These are either transformed into the purest substance of the Sun or else move away toward the heaven. Finally, some are larger and composed of several particles, grooved or otherwise, joined together. These are driven out toward the heaven where (since they are too large to pass through those narrow passages which the globules of the second element leave around themselves), they enter the very places of these globules; and because they have very irregular and branching shapes, they cannot be as easily moved as these globules.

100.   How an aether around the Sun and stars is formed from these
                [latter particles]; and that this aether and those spots are
                included in the third element.

Rather, adhering somewhat to one another, they form there a certain large and very rarefied[101] mass, similar to the air (or rather the aether)

---

[100] That is, the faster-moving material will exert more centrifugal force on the surrounding globules.

[101] By 'rarefied' Descartes seems to mean here that the particles are connected to one another at only a few points, thus forming a sort of net. See Article 102.

which surrounds the earth; this may extend on all sides of the Sun as far as the sphere of Mercury or even beyond. Yet this aether cannot increase indefinitely (although new particles are always coming to it from the dissolution of the spots); because the constant agitation of the globules of the second element through and around it can easily disunite an equal number of others and convert them again into the matter of the first element. And of course we include in the third element all the spots of the Sun and other stars, and also all the aether surrounding them, since its parts are less suited to motion than are the globules of the second element.

101.   That the production and disintegration of spots depend on causes which are very uncertain.

However, the production or disintegration of spots depends upon such minute and uncertain causes that it is not surprising if at times absolutely none appears on the Sun, or if, on the contrary, they sometimes are so numerous that its light is entirely obscured.[102] For if some few of the scrapings of the first element begin to adhere to one another, the beginning of a spot is thereby created; and many other scrapings (which could not [otherwise] adhere to one another unless they lost a part of their agitation by striking against the first scrapings) are subsequently easily joined to it.

102.   How the same spot can cover the entire surface of a star.

It must also be noted that these spots are extremely soft and rarefied bodies when first formed, and therefore easily reduce the impetus of the scrapings of the first element which collide with them, and attach these [scrapings] to themselves. Subsequently, however, their inner surface is not only abraded and thoroughly polished, but also made denser and harder, by the continuous movement of the solar substance to which it is contiguous; while their other surface, which faces the heaven, remains soft and rarefied. Therefore, these spots are not easily disintegrated by the Sun's matter washing against their inner surface unless at the same time it also flows around and over their edges. On the contrary, these spots are constantly increased as long as their edges, rising above the surface of the Sun, do not become denser by contact with its matter. And consequently, it can happen that one and the same spot extends over the entire surface of a star and remains there for a long time before being destroyed.

[102] The French text has, "...that its light is noticeably dimmer." here.

103.   Why the Sun sometimes appears darker and why the apparent
                magnitudes of certain stars are changed.

Thus, certain historians relate that the Sun, for many consecutive days or
even at times for an entire year, has shown a gloomy light which was paler
than usual (like [that of] the Moon), and rayless.[103] And it can be noted
that many stars now appear smaller or larger than they were formerly
described [to be] by Astronomers.[104] There seems to be no reason for this
other than that the light of these stars is {now} dimmed by a greater or
smaller number of spots {than in the past}.

104.   Why some fixed stars disappear or unexpectedly appear.

Indeed, it can even happen that the spots which cover some star become,
{with the passing of time}, so dense as to entirely conceal it from our view:
thus seven Pleiades could formerly be counted, though now we see only
six.[105] On the other hand, it can also happen that a star which we have not
seen before unexpectedly shines forth with a great light in an extremely
short time. {The reason for this is that}, if the whole body of that star has
been entirely covered by a huge thick spot {that will entirely conceal it from
our eyes}, and if at a certain moment the matter of the first element, flowing
toward the star more abundantly than usual, spreads over the outer surface
of the spot; it will completely cover it in a very short space of time, and
make the star look as bright as if it had no spots at all. This star may
continue to show this brilliant light for a long time afterwards, or may lose
it gradually. Thus, toward the end of the year 1572, a star, not previously
seen, appeared in the sign of Cassiopeia, shining with the greatest {very

---

[103] This description of the sun occurs, very nearly verbatim, in Georgius Cedrenus, *Annales
Sive Historiae* . . . (Basel, 1566, p. 304). "Rayless" light is, presumably, light which does not
cast sharp shadows.

[104] The magnitude, or apparent brightness, of a star indicated to which of six different groups
the star belonged; the brightest stars being first magnitude stars, etc. Since, until Descartes's
time, the stars were thought to all be at the same distance from Earth, brightness was taken to
be an indication of size. The stars in a given magnitude differ in brightness, so different
astronomers might well assign the same star to different magnitudes.

[105] The Pleiades is a group of stars located between the constellations Taurus and Aries. In
myth, the Pleiades were said to be the seven daughters of Atlas, but later Greek writers claimed
that one of the original seven had since disappeared.

bright} light, which subsequently grew gradually dimmer, until the star disappeared entirely around the beginning of the year 1574.[106] We also notice other, {more enduring}, stars in the sky which formerly were unknown {to the ancients}.[107] I shall now try to give a fuller explanation of {all} these things.

105.   That there are many pores in the spots through which the grooved particles pass freely.

For example, if star I[108] is entirely covered by the spot defg, this spot cannot be so dense as not to have many pores or {little} passage-ways through which all the matter of the first element, even the grooved particles described above, can pass. For, since it was very soft and very rare at the beginning, many such pores were easily formed in it. And although its parts have become denser {and it has grown harder}; nevertheless the grooved and other particles of the first element, by passing continually through the pores, have prevented these from closing entirely, and have merely allowed them to narrow to a size which is only large enough to allow the grooved particles, {the largest} of the first element, to pass through. And these spaces will be {only as large as is needed to let them pass through from the side at which they usually enter}, such that the channels through which those coming {toward I} from one of the poles are admitted would not be able to admit them if they were to return {from I to the same pole}; nor would they be able to admit those coming from the other pole, because the {spiral} twists {of these particles} run in the opposite way.

106.   How these passages are arranged, and why the grooved particles cannot return through them.

Thus the grooved particles of the first element which {constantly} flow from A toward I (or more precisely, from all that area of the heaven which surrounds pole A, toward the part of the heaven [designated as] HIQ), form for themselves certain passages in the spot defg, along straight lines which are parallel to the axis fd (or converging somewhat on both sides toward d; {because there is more space at A, from which they are coming,

---

[106] The reference is to the "new star" or nova observed by Tycho Brahe.

[107] This is likely the result of the fact that the ancients made no attempt to describe all the stars which are visible to the naked eye. Ptolemy lists about one third of them, for example.

[108] Articles 105–113 refer to Plate XII, Fig. i.

than at I, their destination}}). And the entrances of these passages are distributed over the entire half of its surface [designated as] efg, the exits being on the other half edg, so that the grooved particles coming from A can easily enter through efg and leave by edg, but can neither re-enter at edg nor leave by efg. The reason for this is that the spot consists only of the tiniest scrapings of the first element which, {being very small and having very irregular figures}, form something similar to {a pile of} small branches when they adhere together. So the grooved particles coming from {A through} f {toward d} had to bend {from f} toward d all the extremities of the little branches which they encountered as they passed through the pores. Therefore, if they returned through the same passages from d toward f, the extremities of the little branches, {bent in the wrong direction}, rising up somewhat, would hinder their passage. Similarly, the grooved particles coming from side B have made other passages for themselves {in this spot defg}, the entrances of which are scattered over the entire surface edg, while the exits are on the opposite surface efg.

107.   And why those coming from one pole do not go through the same pores as those coming from the other.

And it must be noted that these pores are hollowed out like snails' shells {and that the thread of these grooves is} in accordance with the shape of the grooved particles which they admit. And accordingly, those pores which are open to certain particles are not open to those others which come from the opposite pole and which are twisted in the opposite direction.

108.   How the matter of the first element flows through these pores.

Accordingly, the matter of the first element can reach star I from both poles through these pores; and because the grooved particles of the first element are bulkier than the others and consequently have more force to proceed along straight lines, they do not usually remain in the star, but having entered through f, immediately leave through d. There, encountering either the globules of the second element or the matter of the first coming from B, they are unable to continue further along straight lines, but, driven back on all sides, return through the surrounding aether xx toward the hemisphere efg. And as many as can enter the pores in the spot or spots which there cover that star again travel through these pores from f to d; and by thus constantly passing through the center of the star and returning through the surrounding aether; they there form a sort of vortex. However, those which cannot be admitted by those pores are either disintegrated by collision with the particles of this aether, or else forced to move away into the heaven through the areas near the equator QH.[109] Of course, it must be noted that the grooved particles which approach star I at each moment are not sufficiently numerous to fill all the passages in spot efg which are hollowed out in a manner suited to them: because even in the heaven, they do not fill all the interstices between the globules of the second element. Rather, a great quantity of more subtle matter must be mingled with them, because of the various movements of these globules; and this more subtle matter would enter those pores with them if the grooved particles, driven back from the other hemisphere of the star, did not have greater force to occupy these pores. Now, all the things which have been said here about the grooved particles entering through hemisphere efg must also be understood about those {coming from pole B} which enter through hemisphere edg. They have hollowed out for themselves, in the star I and the spots surrounding it, other pores, different {and twisted in the opposite direction} from the first ones; and many of them constantly flow through these pores from d toward f. Then, having been driven back on all sides, they return through the aether xx to d; while meanwhile as many are disintegrated or exit near the equator as there are new ones approaching from pole B.

109.   That other pores intersect these crosswise.

As for the rest of the matter of the first element contained within space I: while it revolves around the axis fd, it constantly strives to move away from it; {and to flow through the heaven toward the Equator MY}. For that reason, in the beginning it created for itself certain narrow passages in the spot defg which it has since maintained, and which intersect crosswise those previously mentioned. And some parts of this matter are always flowing through them, because some are also constantly entering through the first pores with the grooved particles. However, inasmuch as all the parts of the spot adhere to one another, circumference defg cannot become sometimes greater and sometimes smaller: and accordingly an equal quantity of the matter of the first element must always be contained in star I.

[109] The equator of the star and the vortex will be a circle at right angles to the page and to the line AB.

110.   That the light of the star can scarcely pass through the spot.

Moreover, for the same reason, that force in which we earlier stated light consists must be completely non-existent, or at least much weakened, in this star. For, inasmuch as its matter is rotated around the axis fd, all the force by which it strives to recede from that axis is deadened by the spot, and does not reach the globules of the second element. Neither can the force by means of which the grooved particles coming from one of the poles tend directly toward the other have any effect here: not only because the grooved particles do not move as rapidly as the rest of the matter of the first element, and are very small in comparison to those of the second {which they would have to drive in order to produce light}; but mainly because those grooved particles which come from one pole do not drive these globules more in one direction than the others coming from the other pole drive them in the opposite direction.

111.   Description of a star suddenly appearing {in the Heavens}.

However, the heavenly matter contained in the vortex {AYBM} which surrounds this star I can meanwhile retain its force {by which it presses against the other surrounding vortices}, although this force may perhaps not be sufficient to excite the sensation of light in our eyes, {for I am assuming this vortex to be very distant}. And it can meanwhile happen that this vortex overcomes the others near it, and presses against them more vigorously than it is pressed by them. From which it would follow that star I would have to increase in size if it were not bounded {on all sides} by the spot defg which impedes this. Thus, if AYBM is the present circumference of the vortex I, we must think that the force by which the globules close to that circumference strive to proceed beyond it and enter the other neighboring vortices is neither greater nor less than, but exactly equal to, the force by which the matter of these other vortices strives to advance toward I. This must be so because only such equality of force could cause that circumference to be where it is {instead of closer to or further from I}. If, however, the force by which, for example, the matter of vortex O tends toward I decreases, without any change occurring in that of the others (and this can happen for several reasons: for example, if its matter flows into {one of the} other {contiguous} vortices, or if the star situated at O becomes covered with spots, etc.); the laws of nature necessarily require that the globules on the circumference Y of vortex I advance beyond it toward P.

This would cause the space in which star I is to increase, if it were not bounded by the spot defg, because all the matter between I and Y also tends toward P. But, because the spot defg does not allow the size of that space to grow, the globules which surround the spot will leave around themselves interstices which are larger than usual; {in order to occupy more space than before. It is, of course, possible for them to move slightly apart in this way, without either separating entirely or ceasing to be attached to the spot, because} the extra matter of the first element, which is contained in these [enlarged] interstices will be so dispersed as to have no great force. But if it happens that {they move so far apart that} either the matter of the first element which collides with these globules as it emerges from the pores of the spot, or any other cause, separates some of them from the surface of the spot; the matter of the first element will immediately fill all the intermediate space, and will have enough force to separate the other nearby globules from that surface. And because the force will increase in proportion to the increase in the number of globules separated from the surface of the spot; it will spread over the whole surface of this spot almost instantaneously. There it will rotate in the same way as the matter inside the spot {which forms star I}, thus driving the surrounding globules of the heaven with as much force as the star I itself would if no surrounding spot impeded its action. And thus star I will {suddenly appear, and} unexpectedly shine with a great light.

112.   Description of a Star slowly disappearing.

Now, if perchance this spot is so thin and rare that the matter of the first element, thus flowing over its exterior surface, can disintegrate it, the star I will not thereafter easily disappear; because in order for that to happen, a new spot covering its entire surface would have to form. However, if the spot is too thick to be thus disintegrated {by the agitation of the matter of the first element}, its exterior surface will grow denser due to the pressure of the matter flowing around it.[110] And if it meanwhile happens that the causes which previously forced the matter of the vortex O to retreat from Y to P are changed in such a way that the matter of vortex I is again repulsed from P toward Y, the quantity of matter of the first element on the surface of the spot defg will decrease, and the surface of the star will be covered by {several other} new spots which will gradually dim its light. If these causes

---

[110] How this material exerts an inward pressure is not explained.

continue, the spots will finally succeed in extinguishing it entirely, and will occupy all the space, formerly filled by the first element, {between the spot defg and the Heaven xx. For the parts of the second element which form the vortex O, advancing from P to Y}; will press all those of vortex I on the exterior circumference APBM more than usual; and thus all those on its interior circumference xx. And those which are thus pressed and are intertwined with the branching particles of that aether which is produced around stars will prevent those grooved particles, and all but the tiniest particles of the matter of the first element, spread over spot defg, from passing {into the Heaven xx} as freely as usual. The grooved and other particles will therefore easily accumulate there and form spots {which, finally occupying all the space between defg and xx, will form a sort of new shell there, over the first one which covers the star I}.

113.   That many pores are hollowed out in all these spots by the grooved particles.

Incidentally, it must be noticed here that the grooved particles hollow out for themselves continuous pores in all these shells of spots, and pass {uninterruptedly} through all, as if through a single spot. For these spots are formed from the very matter of the first element, and therefore in the beginning they are very soft and easily permit these grooved particles to pass through. The same cannot be said of the surrounding aether; although its bulkier particles do indeed retain some vestiges of these pores, since they were created from the disintegration of the spots. However, since these particles conform to the movement of the globules of the second element, they do not always maintain the same situation, and therefore admit the grooved particles proceeding in straight lines only with extreme difficulty.

114.   That the same star can alternately appear and disappear.

Thus, it can easily happen that the same fixed star appears {to} and disappears {from our view} several times; and that each time it disappears, a new shell of spots forms to cover it. For such alternating changes in moving bodies are very common in nature: thus, when a body is driven toward a certain limit by some cause; instead of remaining there, it {generally} proceeds further and then is driven back toward the same limit by another cause. Thus, while a heavy body, suspended from a cord, is moving by the force of its weight down from one side toward the perpendicular, it acquires impetus which carries it beyond that line to the side opposite {that from which it began its movement}; until its weight, surmounting that impetus, causes it to return toward the perpendicular and, thereby, it again acquires a new impetus {which causes it to proceed beyond that same line}. Again, when a vessel is once moved, {even if it has only been moved in one direction}, the liquid in it moves back and forth many times before coming to rest. Now, {because all the vortices which form the Heaven are more or less equal in force}, there exists among the vortices a sort of equilibrium; and whenever the matter of some vortex has departed from that equilibrium, it too may advance and and recoil many times before that motion ceases.

115.   That sometimes an entire vortex, which has a star at its center, may be destroyed.

It can also happen that an entire vortex that contains some such star is absorbed by the other surrounding vortices and that its star, snatched into one of these vortices, becomes a Planet or a Comet. For earlier, we found only two causes which prevent some vortices from being destroyed by others. The first of these, which is that the matter of one vortex is prevented from being able to spread into another by the opposition of its neighboring vortices, cannot apply to all. If, for example, the matter of vortex S^{111} is pressed on both sides by vortices L and N in such a way as to prevent it from spreading further toward D; it cannot, because of its size, be similarly prevented from spreading toward L and N by vortex D, nor indeed by any others unless those are closer to it {than L and N}. Accordingly, this cause does not operate in those vortices which are the closest of all. However, the other cause, which is that that matter of the first element which forms a star in the center of each vortex repulses the surrounding globules of the second from itself toward the other neighboring vortices, does indeed operate in all those vortices whose stars are not enveloped by spots. But there is no doubt that the intervention of denser spots {entirely covering a star} removes this cause, especially when those spots rest upon one another like many shells.

[111] See Plate VI.

148 PART III

116. How a vortex can be destroyed before many spots have accumulated around its star.

From this it is obvious that no vortex is in any danger of being destroyed by other neighboring ones while the star at its center is without spots; but when the star is {entirely} covered and blocked by them, whether the vortex will be absorbed by others sooner rather than later depends solely on its situation in relation to them. Specifically, if it is so situated as to be a great hindrance to the course of {the matter of} other neighboring vortices, it will be destroyed by them before many layers of spots which cover its star can become dense; but if it does not hinder the other vortices so much, it will be diminished only gradually. Meanwhile, the spots besetting its star will grow increasingly dense and will accumulate in ever-increasing numbers {around it}, not only on the outside, {as explained above}, but also within it. Thus, for example, vortex N[112] is so situated that it obviously impedes the movement of S more than does any other vortex neighboring S; for that reason it will easily be carried away by vortex S as soon as its star {at its center} becomes covered with spots. Specifically, the circumference of vortex S, now bounded by the {curved} line OPQ, will then extend to the line ORQ, and all the matter contained between these two lines will approach vortex S and follow its course, while the remainder of the matter, which is between lines ORQ and OMQ, will {similarly} depart into other neighboring vortices. For nothing can preserve vortex N in the situation in which we are now supposing it to be, except the great force of the matter of the first element at its center, which drives the globules of the second element on all sides in such a way that they conform to its pressure rather than to the movement of the other nearby vortices. And as this star becomes covered by spots, this force is weakened and deadened, {and finally disappears entirely}.

117. How there can be very many spots around some star, before its vortex is destroyed.

However, vortex C[113] is situated between the four vortices S, F, G, and H, and the other two, M and N (which we must imagine to be above the first four), in such a way that although {a quantity of} dense spots may accumulate around its star; it still cannot be entirely destroyed, as long as

[112] See Plate VI.
[113] See Plate XII, Fig. ii.

the forces of these six surrounding vortices remain equal. For I am supposing that S and F, and also M which is situated above them at point D, revolve each around its own center, from D toward C. And G and H, and also the sixth one N, which is above them, revolve from E toward C. Finally, vortex C is surrounded by these six in such a way that it touches no others, that its center is equidistant from all of their centers, and that the axis around which it rotates is on the line DE. Thus the movements of these seven vortices are in the greatest harmony, and no matter how many spots there may be blocking the star of vortex C, even if it retains little or no force to carry the surrounding globules of its heaven along in its revolutions, there is no reason for the other six vortices to drive the star from its place, as long as they all remain of equal force.

118.   How these numerous spots are produced.

In order to discover how such a great quantity of spots could have been produced around that star, let us suppose that, in the beginning, its vortex was not smaller than {any} one of the other six which surround it, so that its circumference extended to points 1, 2, 3, 4; and that at its center it had an extremely large star (composed of the matter of the first element which entered from the three vortices S, F, and M through its pole D and headed directly toward C, and from the other three, G, H, and N, through its other pole; and which only left to re-enter those same vortices {through its equator}, at points K and L). Thus this star could have had the force to make all the matter of the heaven {contained within the circumference} 1, 2, 3, 4 rotate with it {and thus form its vortex}. Now because of the inequality and incommensurability of the movements and sizes of the other parts of the universe, {the force of these seven vortices could not remain constantly equal, since} nothing can remain in perpetual equilibrium. Therefore, when vortex C by chance had begun to have less force than its neighbors, some part of its matter passed into them with such violent force that the quantity which passed into them was greater than that required by that difference {between its force and theirs}. Therefore, a portion of the matter of the others {must subsequently have} returned into it, and a similar exchange of matter must have taken place several times.[114]

[114] The relation between the decrease in the rotational force of a star and the increase in the matter which leaves its vortex is not made clear here. Presumably, the decreased force of the star causes the neighboring vortices to expand; and of course this is possible only if they acquire more material from the weakened vortex. See Article 116.

And since, meanwhile, many layers of spots were produced around its star, {in the manner explained above};[115] the force of that star thereby gradually diminished more and more. As a result, the quantity of matter leaving it was {slightly} greater, each time, than that entering it; until the vortex became very small or even until nothing at all remained of it except its star. This star, being surrounded by many spots, can neither mingle with the matter of the other vortices nor be driven from its place by them, as long as these vortices remain equal to one another {in force}. Meanwhile, the spots which surround it must grow increasingly dense; and finally, when one of the surrounding vortices becomes noticeably larger and more forceful than the others (for example, if vortex H increases so much that its surface extends up to line 567), it will easily carry off with it the whole star C, which will no longer be fluid and luminous, but hard and {dark or} opaque, like a Comet or a Planet.

119.   How a fixed Star is transformed into a Planet or a Comet.

Now we must consider how such a hard and opaque sphere composed of an accumulation of many spots must move when it begins to be carried along {in this way} by one of its neighboring vortices. Specifically, it rotates with the matter of this vortex in such a way that it will be driven {by that matter} toward the center of this rotation, as long as it has less agitation[116] than {the parts of} that matter {which surround it}. But the particles of {the matter which forms} a vortex are not all equal, either in speed or in size. Rather, their movement is slower, in proportion to their distance from the circumference, until a certain point; below which they are both smaller and move more rapidly in proportion to their closeness to the center, as was said earlier. Therefore, if this globe is so solid[116] that, before descending to the point at which the parts of the vortex move the most slowly,[117] it acquires a degree of agitation equal to that of those parts among which it is located; it descends no further, and will proceed into other vortices, and become a Comet. On the other hand, if it is not sufficiently solid to acquire so much agitation, and therefore descends below that point {at which the parts of the

---

[115] See Article 112.

[116] See Articles 121 and 122 for Descartes's discussion of the meanings of 'agitation' and 'solidity'.

[117] This will be the point at which the effect of the central star's rotation on the material of the vortex is smallest, and will indicate the largest orbit possible for a planet in that vortex.

vortex move the most slowly}, it will remain a certain distance from the star which occupies the center of this vortex; and will become a Planet revolving around it.

120.   Where such a Star is transported when it first ceases to be fixed.

Let us suppose, for example, that the matter of the vortex AEIO[118] is now forcibly beginning to carry star N along with it, and let us see in what direction this matter will carry it. Since all this matter revolves around the center S, it certainly must also strive to recede from it, as I explained earlier. Consequently, there is no doubt that the matter currently situated at O will, by rotating through R to Q, drive the star in a straight line {from N} toward S, {and thus cause it to descend in that direction}.[119] For, when we explain, later on, the nature of weight, we will understand that such a movement of star N (or any other body) toward the center of the vortex in which it is located can be {properly} called its descent. Thus N must, I say, be driven in the beginning, for we do not yet understand it to have any other movement. But this matter, by immediately also flowing around N or all sides, carries N with it in a circular movement from N toward A. Since this circular movement gives the star the force to recede from the center S, {and since these two forces are opposed}; the degree of its descent depends solely on its solidity. If it has very little solidity, it will descend a great deal toward S: and if it has greater solidity, it recedes from S.

121.   What we understand by the solidity of bodies and by their agitation.

What I understand here, by the solidity of this star, is the quantity of the

---

[118] See Plate VI.

[119] Descartes seems to be claiming here that the material which is rotating further from S than N is somehow drives N toward S, though he gives no explanation of this. In Article 121, however, Descartes claims that the star is driven toward S as a result of the matter between N and S having a greater centrifugal force than does N. Thus, this material will flow past N and drive it toward S, somewhat in the way that the water above a light, immersed body will displace that body upwards until it floats. In a rotating fluid, the less dense parts (and solid objects less dense than the fluid) do move toward the axis of rotation, but this effect is solely dependent on a difference in density. Descartes takes it to depend on the relative solidity of the material concerned, but see note 121.

matter of the third element,[120] of which the spots surrounding it are composed, in proportion to its volume and surface area. For in fact, the force by which the vortex AEIO carries N circularly around the center S must be judged by the size of the surface [of the star] which it encounters; because the larger this surface is, the greater the quantity of matter acting against the surface. However, the force with which the same matter drives N {down} toward S must be estimated by the volume of N. For although all the matter in the vortex AEIO strives to move away from S, not all of this matter acts on star N, but only that part of it which rises {to take the star's place} when it descends; and this is equal in volume to the space N occupied. Finally, the force which this star acquires from its motion {around the center S with the matter of the Heaven} to continue {to be thus transported or} to thus move, which I call its agitation; must be estimated neither by the size of its surface area nor by the total quantity of matter {which composes it}, but only by the quantity of the matter of the third element, the particles of which adhere closely to one another and form the spots enveloping it.[121] As for the matter of the first or even the second element, it is continually leaving this star and being replaced by new matter. Consequently, this new matter approaching cannot retain the force of

---

[120] This seems to be as close as Descartes ever comes to something like the concept of mass. If "quantity of matter of the third element" is taken in this way, then the ratio between quantity of matter and total volume would be analogous to an object's density. It must be noted, however, that Descartes takes the "quantity of matter" of the third element to be determined by its volume rather than by its inertia.

[121] In fact, Descartes has three different concepts of solidity here. In general, he intends the solidity of a body to be a determining factor in its tendency to move in certain ways and to be moved in other ways. Three such tendencies are involved here, giving rise to three different concepts. First, there is the tendency of the body to be carried along by the fluid matter of the vortex; and, for any given velocity, this will be proportional to the ratio of the body's quantity of third-element matter to its surface area. (The text does not actually say this, but unless this is what is meant, it is difficult to make sense of the first sentence of this article. Further, Descartes's final example in Article 122 bears this out, and the French text clearly takes this to be his intention; see, e.g., Article 125.) Second, there is the body's tendency to recede from the center (or its ability to resist being driven toward the center). This will be a function of something like its density, that is, of the proportion of its total volume which is occupied by the matter of the third element. Finally, there is the tendency which the body has to continue its motion in a straight line, or, as Descartes here calls it, its agitation. This will depend, among other things, on the body's "quantity of motion", in the language of Part II, and thus, in Descartes's view, on the total quantity of the matter of the third element which the body contains. Thus, solidity in this third sense will determine, for a given speed, the force which a body has to retain its motion in a straight line.

agitation acquired by the matter which has already left, which, in any case, was very small. And the motion which the new matter possessed from other sources was only a determination to move in a certain direction {rather than in others}; which determination can continually be changed by various causes.

122.   That solidity does not depend on matter alone, but also on size and shape.

Thus, here on earth, we see that, once moved, gold, lead, or other metals retain more agitation, or force to continue in their movement, than do pieces of wood or rocks of the same size and shape; and consequently metals are also thought to be more solid, or to contain more matter of the third element and smaller pores filled with the matter of the first and second.[122] But a small sphere of gold can be so tiny that it will not have as much force to retain the movement communicated to it as will a much larger sphere of rock or wood. And a lump of gold can also assume shapes such that a wooden sphere of smaller size can be capable of greater agitation; for example if it is drawn out into threads or {forged} into thin plates or hollowed out with numerous holes like a sponge, or if it in any other way acquires more surface area, in proportion to its matter and volume, than that wooden sphere.

123.   How the celestial globules can be more solid than an entire star.

Thus, it can happen that star N has less solidity, or less ability to continue its movement, than the globules of the second element which surround it; even though it may be very large and covered with fairly many layers of spots. For these globules, in proportion to their size, are as solid as any body can be, because we understand that they contain no pores filled with other ... matter;[123] and because their figure is spherical; the sphere being the figure which has the least surface area in proportion to its volume, as

[122] The third sense of 'solid' seems to be intended here, although if the bodies are of the same size, the proportion of matter of the third element to the total volume would also be greater for the object with more solidity.
[123] The Latin text says "other more solid matter," which does not seem to make sense. 'More solid' is omitted from the French text; 'less solid' was probably intended.

Geometers know.[124] Furthermore, although there is a very great disparity between their tininess and the magnitude of a star, this is in part compensated for by the fact that it is the force of many of these globules taken together, rather than individually, which opposes the force of this star. So that, while they are rotating with some star around the center S, and they, as well as the star, are all striving to recede from that center; if it happens that this force[125] in the star is greater than the united forces of all the little globules which are required to fill its place, the star must move away from the center and cause these globules to descend into its former place. On the other hand, if the star has less force, it will be driven by the globules toward S.

124.   And how these [globules] can also be less solid.

Further, it can also easily happen that star N has much more force to continue in its movement along straight lines than the globules of heavenly matter surrounding it. This can occur even though the star contains less of the matter of the third element than the number of globules necessary to fill a volume equal to that of the star contains of the second [element].[126] Because these globules are separated from one another and have various {individual} movements; although their united forces act against the star, they cannot all unite their force simultaneously in such a way [as to ensure] that no part of their force is wasted. In contrast, all the matter of the third element, comprising the spots enveloping this star and the aether surrounding it, forms one single mass which is moved together as a whole, and thus all the force which it has to continue its motion is applied in a single direction. For a similar reason, fragments of ice or pieces of wood, which are floating on the surface of a river, can be seen to pursue their courses in

[124] Since the spherical globules are completely filled with second element matter, the ratio of the quantity of that matter to their surface will be greater than for any other shape. Further, the ratio of the quantity of that matter to total volume will be one, and the total quantity of second element matter will be greater than in any other kind of body of the same size. Thus, each globule will have as much solidity, in all senses of that term, as its size permits, and its solidity may be determined simply by considering its size. This suggests that for globules of a given size, 'solidity' may be taken in any of its senses indifferently, as Descartes seems to do here.

[125] This seems to be the force resulting from solidity in the second sense.

[126] Solidity in the third sense is involved here, and in this sense a star will always be less solid than an equal volume of globules, since the star does not consist entirely of matter of the third element.

straight lines with greater force than does the water itself; and therefore, they are accustomed to strike the curves of the bank much harder, although they contain less of the matter of the third element than an equal volume of water.[127]

125.   How some [globules] are more solid than a given star, and others less.

Finally, it can happen that the same star may be less solid than certain globules of the heaven, and more solid than some other rather smaller ones, both for the reason just stated, {namely, that the forces of several globules are less unified than those of one larger body equal to all of them}, and also because, although the quantity of the matter of the second element in all the globules which occupy a given [amount of] space[128] may be the same whether they are {very} small or {quite} large, the smaller ones have {less force, because they have} more surface area {in proportion to the quantity of their matter}; and therefore they can be drawn off their course and turned aside in other directions more easily than the larger ones, either by the matter of the first element filling the spaces which they leave around themselves, or by any other bodies {which they encounter}.

126.   Concerning the origin of the movement of Comets.

If, therefore, we now suppose the star N[129] to be more solid than the globules of the second element which are fairly distant from the center S and which we are supposing to be equal in size to one another; it will be possible in the beginning for it to be transported in various directions, and to approach S more or less, depending on the various arrangements of the other vortices whose vicinity it is leaving: for these can either restrain it or drive it in various ways. Its motion also depends on its proportionate solidity, because the greater this is, the more that will prevent other causes from later turning it from its original course. Nevertheless, the nearby vortices certainly cannot drive it with extremely great force, since we are supposing it to have been previously at rest relative to them; and therefore, it cannot move {in a direction} contrary to the motion of vortex AEIO,

---

[127] Presumably, this is shown by the fact that they float.
[128] The French has "...fill a space equal to that of the star..." here.
[129] Articles 126 through 132 refer to Plate VI.

toward those areas [of the vortex] which are between {the side} I {O of its circumference} and {the center} S. Rather, it will move only {in the opposite direction}, toward those areas between A and S. And it must finally arrive at some point {in that area} where the line described by its movement will be tangent to one of the circular lines described by the heavenly globules as they revolve around the center S. From there, it will pursue its subsequent course in such a way as to move constantly further away from S, until it leaves the vortex AEIO, and enters another. Thus, if in the beginning it moves along the line NC; when it has arrived at C (where this curved line NC is tangent to the circle described there by the globules of the second element {as they revolve} around S), it must at once recede from S along the curved line C2 which passes between this circle and the straight line tangent to it at point C. For this star, having been carried to C by the matter of the second element which was further from S than that which is at C, and which consequently was moving more rapidly, and being, we suppose, more solid than the matter [at C]; cannot fail to have greater force to continue its movement along the straight line tangent to that circle. However as soon as it has receded from point C, it encounters matter of the second element which is moving {a little} more rapidly {than that at C} and which causes the star to deviate somewhat from the straight line. And this matter, by increasing the star's speed, causes it to rise further along the curved line C2; the deviation of which from the straight line tangent [to the circle] diminishes in proportion to the solidity of the star and the speed at which it moved from N to C.

127.   Concerning the continuation of the movement of Comets through diverse vortices.

While it is thus proceeding through this vortex AEIO, it acquires so much agitation as to have the force to pass into other vortices from which it {subsequently} proceeds into still others. {In this way, it continues its movement, concerning which two things must be noticed here. The first is that, when this star passes from one vortex into another, it always drives before it some portion of the matter of the one it is leaving, and cannot be entirely free of it, until it is well within the limits of the other. For example}, when it leaves vortex AEIO, and is at 2, it still has {some of} the matter of this vortex revolving around it, and cannot be entirely free of this matter until reaching 3, in vortex AEV. And, similarly, it takes with it the matter of this second vortex to 4, within the limits of the third vortex; and it carries matter of this third to 8, within the fourth; and this always occurs whenever it migrates from one vortex to another. {The second thing which must be noticed is that the course of} this star describes a line whose curvature varies according to the diverse movements of the vortices through which it passes. Thus, part 234 of this line is quite differently curved from the previous part NC2; because the matter of vortex F revolves from A through E to V, and that of vortex S from A to E to I. And part 5678 of that line is almost straight, because the matter of the vortex in which it is situated is assumed to rotate on the axis XX. The stars which migrate in this way from one vortex to another are {those which we call} Comets, and I shall now try to explain all their phenomena.

128.   What the principal phenomena of Comets are.[130]

First of all, it is observed that Comets pass, one through one region of the heaven, and another through a different region, without following any rule known to us.[131] They vanish from our sight within a few days or months; and they never cross more than half of the [sphere of the] heaven, or certainly not much more, and frequently much less. Further, when they first become visible, they usually appear fairly large and do not subsequently increase in size, except when they cross a very large part of the heaven; however, they always gradually diminish toward the end {of their appearance}. At the beginning, or around the beginning of their movement,

---

[130] These "phenomena" are highly suspect. The apparent motions of comets are quite irregular and very difficult to analyze. The first successful determination of a comet's path was made by Dörfel in 1681. Their apparent size and speed vary greatly with the relative positions of the Earth and the comet. Visible comets pass very close to, and frequently intersect, the Earth's orbit; but the Earth may or may not be near the points of intersection at the time. Descartes seems to be generalizing here from only a few observations, although he admits that the comet of 1475 is atypical.

[131] Comets obey three rules. They move in conic sections with the sun at one focus, and they obey Kepler's second and third laws of planetary motion. This was not known in Descartes's time, of course. Descartes probably has in mind here the fact that the paths of comets bear no particular relation to the ecliptic, and that their speed varies widely. Descartes does not seem to have known that some comets move in a direction opposite to the planetary motions, an impossibility on his view. However, Regiomontanus' description of the comet of 1475 specifically states that the comet's motion was opposite to the direction of revolution of the planets; and Descartes's letter to Du Puy leaves no doubt that Descartes had read the description in its entirety: see note 136. Of course, the *Principles* had already been published when Descartes saw the complete description; but the portion of the description quoted in the *Libra Astronomica* clearly indicates, though it does not explicitly state, the comet's direction.

they appear to move very rapidly, though very slowly toward the end. I remember having read of only one* which travelled across approximately one half of the heaven. It was said to have appeared in the year 1475 among the stars of Virgo, having at first a small body and slow movement. Shortly afterwards, having reached astonishing size, it crossed the North Pole so rapidly that it described thirty or forty degrees of {one of} the great circles {which we imagine on the sphere} in a single day. And it finally vanished from sight near the stars of Pisces or in the sign of Aries.[132]

* [Descartes's own footnote] In the *Libra Astronomica* of Lotharius Sarsius or Horatio Grassius,[133] where it is spoken of as two comets. But I judge it to have been a single one, an account of which is given by two authors, Regiomontanus and Pontanus.

129.   The explanation of these phenomena.

Now, all these things {which have been observed} can be {very} easily understood. For we see that the same Comet {which we described} traverses one part of the heaven in vortex F and another in vortex Y, and that there is no part [of a vortex] which it cannot at some time cross in this way.[134] We must think, too, that it always maintains approximately the same speed, namely that which it acquires as it passes through the extremities of these vortices, where the matter of the heaven is so rapidly moved that it completes its entire revolution in a few months, as has already been said.[135] Consequently, the Comet, which completes only about half of such a revolution in vortex Y, and much less in vortex F, and which can never complete much more than half in any, can remain in the same vortex only a few months. Then, if we consider that it is only visible to us while {in the first Heaven, that is}, in the vortex near the center of which we live, and that it cannot be visible there until it ceases to be surrounded and followed by the matter of the vortex from which it is coming: we shall be able to understand why, although a given Comet always moves at approximately the same speed and remains the same size, it must nevertheless appear to be larger

[132] In Descartes's time, the constellation Pisces was in the sign of Aries.
[133] This work was published in 1619 as a reply to Galileo's *Discourse on the Comets*; itself a reply to a pamphlet of Grassius' on the comet of 1618 in which it is argued (quite correctly) that the comet was non-terrestrial. The quotation therein of Regiomontanus' description of the comet of 1475 is roughly as Descartes gives it in this article. The description of one, or possibly two, comets quoted from Pontanus are excerpts from a (very bad) poem which is so vague as to be quite uninformative. See A. & T., IV, 150–152, and 665.
[134] That is, its path will not bear any special relationship to the equator of any given vortex; though for a given comet and a given vortex, there will be only one path, of course.
[135] See Article 82.

and more rapid at the beginning of its appearance than at the end, and must appear largest and most rapid in the middle.[136] For if we think that the spectator's eye is near the center {of vortex} F; the Comet will appear to him much larger and more rapid [when it is] at 3, where it will first begin to be visible, than at 4, where it will cease to be visible: because the line F3 is much shorter than the line F4, and the angle F43 is more acute than the angle F34.[137] If, however, the spectator is at Y, this Comet will certainly appear somewhat larger and more rapid to him when it is at 5, where it will become visible to him, than when it is at 8, where he will lose sight of it; but it will appear to him largest and most rapid while it is between 6 and 7, where it will be closest to the spectator. So that, {if we take vortex Y to be the first Heaven, in which we are}, it will appear among the stars of Virgo, when it is at 5; and near the North pole as it passes between 6 and 7, and there during a single day it can cover thirty of forty degrees, and finally disappear at 8, near the stars of Pisces: in the same way as this wondrous Comet of the year 1475, said to have been observed by Regiomontanus.

130.   How the light of the fixed stars reaches the Earth.

It is true that one may ask here why Comets are not visible unless they are in our heaven, while fixed Stars are clearly visible, although very far beyond it. But the difference lies in the fact that the fixed Stars, emitting their own light, cast it with a much stronger {and more intense} agitation than do Comets, which only reflect light to us from the Sun. And if we notice that the light of each star consists in that action by which all the matter of the vortex in which it is situated strives to recede from it along straight lines from all points on its surface; and thus presses the matter of the surrounding vortices along the same straight lines or others equally effective (that is, those which {the laws of} refraction {cause them to} produce, when they pass obliquely from one body to another, as I explained in the *Dioptrics*);[138] it can easily be believed that the light of the

---

[136] This is in accord with Regiomontanus' description mentioned in Article 128, at least as far as its apparent speed is concerned. In a letter to Du Puy of January, 1645, thinking him for providing a complete version of the description excerpted in the *Libra Astronomica*, Descartes writes: "I am now curious about only one remaining point, namely, the size of this comet; for according to my reasoning, it must have appeared so noticeably larger in the middle of its course, than at the beginning and the end, that it is unbelievable that Regiomontanus did not mention the fact." A. & T., IV, 152.

[137] That is, when at 3, its motion will be more nearly at right angles to the line of sight.

[138] See *Dioptrics*: Second Discourse.

stars, not only that of those closest {to earth}, like f, F, {L, and D} (for I am supposing the earth to be not far from S); but also that of the more distant ones like Y, has the force to affect the eyes of the inhabitants of the earth. For, inasmuch as the forces of all these stars {among which I am including the Sun}, taken together with those of the vortices surrounding them, are always in equilibrium; the force of the rays {of light} coming from F toward S certainly is diminished by the opposition of the matter of vortex AEIO; but it is not entirely diminished except at the center S. Therefore, many rays can reach the earth, which is a little distance from that center. Similarly, the rays coming from Y to the earth do not lose any of their force as they pass through vortex AEV, except because of the distance; because the matter of vortex AEV does not diminish their force more as it strives to recede from F toward the part VX of its circumference, than it increases their force as it also strives to move from F toward the part AE of that circumference.[139] And the same is to be understood of the other stars.

131.   Whether the fixed Stars are seen in their true locations; and what the Firmament is.

We must also notice here that the rays coming from Y toward the earth obliquely intersect the lines AE and VX (which represent the surfaces separating these vortices {S, F, and Y} from one another); and that consequently these rays must be refracted there.[140] It follows from this that, from the earth, the fixed stars do not all appear to be in the places in which they are truly situated, but {that we see them} as if they were at the points on the surface of {our} vortex AEIO through which pass those of

[139] Descartes seems to be suggesting here that light consists of rays which travel from a star to the eye, which is not in accord with what is said in Articles 63 and 64. Of course, 'ray' may not refer to anything physical here, but merely to an instantaneously transmitted pressure. In a letter of Aug. 22, 1634, possibly to Beeckman, Descartes writes: "I said recently, when we were together, not that light was instantaneously moved, as you write, but . . . that light reaches our eyes instantaneously from luminous bodies, and I also added that this was to me so certain that if its falseness could be proved, I would be prepared to acknowledge that I know nothing about Philosophy." Descartes then claims that unless light reaches us instantly, celestial events would not occur at their predicted times.
[140] This would be true only if different vortices had different indices of refraction, and Descartes gives no reason to suppose this here. He probably intends it to follow from the fact that light is weakened or impeded when it enters a vortex other than its own, since in the Dioptrics, he explains refraction as the result of a difference between two media in their ability to transmit light. Cf. Article 132.

their rays which reach the earth, or the neighborhood of the Sun. And it is thus possible that the same star appears to us to be in two or more places, {and that we therefore count it as several. For example, the rays of Star Y can equally well reach S by passing obliquely via the surface of vortex f, as by passing via that of the one marked F, with the result that we must see this Star in two places, i.e., between E and I and between A and E}. Inasmuch as the places in which the stars are {thus visible remain constant, and} appear not to have changed during the time Astronomers have been observing them, it seems to me that the Firmament is to be understood as nothing other than these surfaces {which separate the vortices from one another, and which cannot be altered without the apparent positions of the Stars changing too}.

132.　　　Why we do not see Comets when they are outside our heaven; and, incidentally, why coals are black and ashes white.

As for the light of Comets: inasmuch as it is much weaker than that of the fixed Stars, it does not have enough force to affect our eyes except when they subtend a wide enough angle; therefore, their distance prevents us from seeing Comets when they are too far from our heaven: for it is known that the angle subtended by a body diminishes in proportion to its distance from us. However, when Comets draw nearer to our heaven, there can be various reasons why they are not visible when they first enter our heaven; although it is not easy to determine which of these causes is the most important.[141] For example, if the spectator's eye is at F, he {will not begin to see the Comet illustrated here until it is at 3, and} will not yet see it when it is at 2, because it will still be surrounded by the matter of the vortex it has just left, {as has been explained above}; but he will be able to see it when it is at 4, although the distance {between F and 4} is greater {than that between F and 2}. The reason for this may be the way in which the rays of star F, travelling toward 2, are there refracted on the convex surface of the matter of vortex AEIO, which still surrounds the Comet. For this refraction deflects the rays from the perpendicular, in the way that I explained in the Dioptrics, i.e., because they have much more difficulty in passing through

[141] The actual cause, of course, is that they become visible only when they are quite close to the Earth. On Descartes's view, however, a comet that came as close as Saturn would become a planet. See Article 140.

the matter of the vortex AEIO than through that of the vortex AEVX.[142] Thus, far fewer reach the Comet than would be the case if this refraction did not occur; and those fewer rays which it reflects toward the eye can be too weak to make it visible. However, another extremely likely reason is that, just as the same side of the Moon always faces the earth, so each Comet has one side which it always turns toward the center of the vortex in which it is situated, and that only this side is suited to reflect the rays. Thus, when the Comet is at 2, it still has the side suited to the reflection of light turned toward S, and so cannot be seen by those who are at F; but, proceeding from 2 to 3, in a short time it turns {this side} toward F, and therefore then begins to be visible. For it is entirely in conformity with reason for us to think, first, that while the Comet passes from N through C to 2, the side turned toward the star S becomes more agitated and rarefied by the action of that star than does its opposite side. And, second, it is reasonable to think that the slenderer and, so to speak, softer particles of the third element which are on {that side of} the Comet's surface are separated from it by this agitation; which makes that side more suited to the reflection of rays than the other. In the same way, it will be possible to understand from what is said further on about fire, that the reason why extinguished coals appear black is simply that all their surfaces, internal as well as external, have been covered by these softer particles of the third element. And when these softer particles are separated from the rest by the [subsequent] force of fire, the black coals are transformed into ashes, which are composed solely of hard and solid particles and are therefore white. And no bodies are better suited to the reflection of rays than white ones, while none are less suited to this than black ones. Third, there is good reason for us to think that the more rarefied part of the Comet is less suited to motion than the other, {because it is less solid}. Therefore, according to the laws of Mechanics, it must always be in the concave part of the curved line which the moving Comet describes, because that side will thereby proceed slightly more slowly than the other. Since the concave part of its path always faces the center of the vortex in which the Comet is (as here the concave part NC2 faces the center S, the concave part 234 faces F, etc.): the Comet is, therefore, turned in passing from one vortex to another. Similarly, we see that when arrows fly through the air, their feathered {lightest} part is always lowermost when they are rising and uppermost when they are descending.

[142] Presumably, the matter of the vortex which the comet carries with it still retains some of its own motion and thus resists the light of star F.

Finally, it is possible to give many other reasons why Comets are only seen by us while they are passing through our heaven; because the suitability of a body for the reflection of light is altered by the slightest changes. Concerning effects of this kind, of which we do not have sufficient experience, it must suffice to give probable causes, even though these may perhaps not be the true ones.

133.   Of the tails[143] of Comets and their various phenomena.

In addition to these things, however, a sort of long hair of rays [of light] is observed to shine around Comets, and from this they took their name. Moreover, this tail is always seen on the part more or less turned away from the Sun: so that if the Earth is situated on a straight line between the Comet and the Sun, the tail is seen surrounding the Comet on all sides. And the Comet of the year 1475, when first seen, was preceded by a tail; toward the end of its appearance, however, because it was situated in the opposite part of the heaven, it was followed by a tail.[144] This tail also varies in length; sometimes because of the {apparent} size of the Comet, for none appears in the smaller ones, or, indeed, in large ones which, receding from our view, appears very small. Sometimes it varies on account of position; for, all else being equal, the further the earth is from a straight line which can be drawn from the Comet to the Sun, the longer the Comet's tail is.[145] And at times, when the Comet is hidden beneath the Sun's rays, only the extremity of its tail is seen, resembling a beam of fire. Finally, this tail sometimes varies slightly in width; it is sometimes straight and sometimes curved;[146] and it may or may not be turned directly away from the Sun.

[143] Literally, 'hair', although 'tail' is now the customary term; 'comet' is from the Greek term meaning "long-haired".
[144] Since a comet's tail points away from the sun, it will be followed by the tail as it approaches the sun and preceded by it as it recedes. Descartes's ground for this description is undoubtedly the poem of Pontanus, mentioned in connection with Article 128, which does make this claim. A comet's apparent motion is not its true motion, of course; and the descriptions relied upon by Descartes contain a serious source of error. If one regards the Earth as motionless, as any observer in 1475 would have, then, as the comet approaches the sun, the tail can *appear* to precede it at first and then to turn and follow it as the comet moves toward the sun more rapidly. Of course, Descartes may have thought that due to a comet's extreme distance and speed, the observational effects of the Earth's motion would be negligible.
[145] That is, since the tail always points away from the sun, it will appear shorter when viewed at an angle than when seen directly from the side.
[146] Since the material of the tail is further from the sun than the body of the comet is, it will move more slowly; obeying Kepler's third law. Thus, the tail will always be curved; whether it appears curved depends on the viewer's position.

134.   Concerning a certain refraction, on which this tail depends.

In order that the reasons for all these things may be understood, it is necessary to consider here a certain new kind of refraction which was not discussed in the *Dioptrics*, because it is not observed in terrestrial bodies. From the fact that the heavenly globules are not all equal to one another but gradually become smaller from a certain boundary-line (within which is contained the sphere of Saturn), onward to the Sun; it certainly follows that, when rays of light which are transmitted by the larger of these particles reach the smaller ones; they must not only progress along straight lines, but must also be partly refracted and dispersed [from these straight lines] to both sides.

135.   The explanation of this refraction.

As an example, let us consider this figure,[147] in which some fairly large balls are resting on many very much smaller ones, and let us suppose them all to be in continual motion, like the globules of the second element described before. If one of them is driven in a certain direction; for example, if ball A is driven toward B, its action is immediately communicated to all the others situated on a straight line drawn from it to B. Here it must be noticed that this entire action reaches C from A, but that only a certain part of it can pass from C to B, while the rest is dispersed toward D and E. For ball C cannot drive the small ball 2 to B without also driving the two others, 1 and 3, toward D and E, {thereby also driving all those contained in the triangle DCE}. However, the situation is not comparable when ball A drives balls 4 and 5 toward C. For its action {by which it drives them}, though received by those two balls 4 and 5 in such a way that it also seems to be deflected toward D and E, nevertheless proceeds directly toward C. This is both because these balls 4 and 5, being equally supported on both sides by those near to them, return all that action to ball 6; and also because their continuous motion prevents this action from ever being received simultaneously by two balls, in any given period of time, and only allows it to be transmitted successively, first by one, {which is disposed to turn it aside in some direction}, and then by the other, {which is disposed to turn it aside in the opposite direction, with the result that it always continues along the same straight line}. But when ball C

[147] See Plate XIII.

drives the three {smaller} balls 1, 2, and 3 simultaneously toward B, its action cannot thus be transmitted by them to a single ball. For however much they are being moved, some of them always receive this action obliquely; and for that reason, although they {always} transmit the principal ray of this action directly toward B, they nonetheless disperse innumerable other weaker ones on both sides toward D and E. In the same way, if ball F is driven toward G; when its action reaches H, it is there communicated to small balls 7, 8, 9, which do indeed send its principal ray to G, but also disperse others toward D and B. But here we must notice the difference made by the degree of obliquity of the incidence of their action on the circle CH: for the action coming from A to C {sends its principal ray toward B and} disperses rays equally on both sides toward D and E, because the line AC intersects this circle at right angles. However, since the action coming from F to H intersects the same circle obliquely, it is dispersed only [on the side of the line which is] toward the center of the circle; at least if we suppose the angle of incidence to be ninety degrees.[148] However, if we suppose this angle to be less, some of the rays of this action will also be dispersed in the other direction; although these will be much weaker than the former. Thus, they will scarcely be perceptible, except when the angle of obliquity is very small. On the other hand, those rays which are obliquely dispersed toward the center of the circle are stronger the greater the angle of obliquity [of the principal ray] is.

136.   The explanation of the appearance of the tail.

Once the demonstration of all these things has been grasped, it is easy to apply it to the heavenly globules; for although there is no place in which the larger of these globules touch other much smaller ones {in this way}; yet because these gradually decrease in size from a certain boundary-line onward to the Sun, as has been said; it can easily be believed that the difference between those which are beyond Saturn's orbit and those which are close to the Earth's orbit is no less great than that between the large and small balls just described. And thence it can be understood that the effect of this inequality in [the region of] this orbit of the Earth must not be different than if the smallest globules immediately followed the largest ones; {the only difference is that in the latter case, the rays of this action are greatly deflected at only one point, while in the case of the Heavenly globules which

---

[148] That is, if FH is tangent to the circle.

grow successively smaller, the deflection is only gradual}. Nor must it be different in intermediate places, except that the lines along which these rays are dispersed will not be straight but slightly curved. Specifically, if S is the Sun,[149] 2345 is the orbit around which the Earth is transported in the space of a year, following the order of the markers 234, if DEFG is that boundary line below which the heavenly globules begin to grow gradually smaller and smaller until they reach the Sun (as we stated above, this boundary-line does not have the form of a perfect sphere, but of an irregular spheroid, much flatter near the poles than at the ecliptic), and if C is a Comet situated {beyond Saturn} in our heaven; it must be thought that the Sun's rays which strike this Comet are so reflected toward all areas of the spheroid DEFGH that those which intersect it perpendicularly at F for the most part continue directly to 3, though some are also variously dispersed {from that line}. Those which intersect it obliquely at G not only continue directly toward 4, but are also to some extent refracted toward 3; and finally, those which intersect it at H do not reach the Earth's orbit directly at all, but [reach it] only because they are {deflected}[150] toward 4 and 5; and so on. From this it is obvious that, if the Earth is in area 3 of its orbit, this Comet will be seen from the Earth with its tail extending on all sides; this type of Comet is called a Rose: for the rays which travel directly from C to 3 appear as its body while the other weaker ones which are {deflected} from E and G toward 3 appear as its tail. However, if the Earth is at 4, the [body of the] same Comet will be seen by means of the straight rays CG4, and its hair, or rather its tail, extended in only one direction, will be seen by means of the rays which (coming from H and other points between G and H) are {deflected} toward 4. Similarly, if the Earth is at 2, the Comet will be seen from it by means of the straight rays CE2, and its tail by means of the oblique ones which are between CE2 and CD2; and there will be no other difference, except that the Comet will be seen in the morning if the eye is at 2, and will have its tail preceding it; while if the eye is at 4 the Comet will be seen in the evening with its tail following it.[151]

---

[149] See Plate XIV.
[150] the Latin has 'reflected' here; presumably, "reflected from the Comet" is meant.
[151] This will be true of some comets, depending on their direction of motion, but it will not be true of all.

137.   How beams also appear.

Finally, if the eye is near point {5,[152] it is obvious that} we shall be unable
to see the Comet itself, because the Sun's rays will prevent this {since the
Sun will be between us and it}. We shall be able to see only a part of its tail,
which will resemble a beam of fire and will appear either in the evening or in
the morning, depending on whether the eye is nearer point 4 or point 2; so
that, if it is precisely at mid-point 5, it is possible that this same Comet will
make visible to us two beams of fire, one in the evening and the other in the
morning, {by means of the curved rays coming from H and D toward 5. I
say [only] that this is possible because, unless the Comet is very large, its
curved rays will not be sufficiently strong to be perceived by our eyes}.

138.   Why the tail of Comets is not always seen in the area directly
            opposite the Sun and does not always appear straight.

And in fact, this hair or tail {of a Comet} must sometimes appear
{exactly} straight, and sometimes slightly curved; and must sometimes be
on the straight line passing through the centers of the Sun and the Comet,
and sometimes deviate slightly from it; finally, it must vary in width, or
even in brightness; i.e., when the lateral rays converge toward the
[observer's] eye. For all these variations follow from the irregularity of the
spheroid DEFGH: since near the poles, where its figure is flatter than
elsewhere, it must make the tails of Comets appear straighter and wider;
but in the curve between the poles and the ecliptic, they must appear more
curved and turned {somewhat} away from the side directly opposite the
Sun; finally, along this curved line, they must appear brighter and
narrower. And I do not think that anything has so far been observed about
Comets, except what must be considered a fable or a miracle, whose cause
has not been included here.

139.   Why the fixed Stars and the Planets do not have similar tails.

The only {remaining} question is why tails do not also appear around the
fixed stars or around the higher planets, Saturn and Jupiter, {in the same
way as around Comets}. But it is easy to answer this question. First of all,
even Comets are not usually accompanied by such tails when their
apparent diameter is not greater than that of the fixed stars; because then

[152] The Latin has "point S" here; the mistake is corrected in the French.

the secondary rays {which form the tail} are not sufficiently strong to affect
our eyes. Then, with regard to the fixed stars, {we must notice specifically
that} (inasmuch as they shine by their own light and not by light borrowed
from the Sun) if some tail appeared around them, it would have to be
{equally} distributed on all sides, and consequently very short, {like that of
the Comets we call Rose Comets}. In fact we do see such hair around them;
for their figure has no uniform outline, and they appear surrounded by
indistinct rays on all sides; this may also be why their light is so glittering
{or twinkling}, although many other reasons for this can be given.
However, as far as Jupiter and Saturn are concerned, I do not doubt that
they too sometimes appear with a short tail on the side away from the Sun,
{in countries} where the air is {very clear and} very pure. And I well
remember having read somewhere that this was observed in the past,
though I do not recall the author's name. As for Aristotle's remark, in the
first book of his *Meteorology*, Chapter 6, that the Egyptians sometimes
perceived such hair around the fixed Stars, I think it should probably be
taken to apply to the planets instead. And the hair which he himself claims
to have seen around one of the stars in the thigh of Canis must have been
the result of some very oblique refraction in the air, or, more probably, of
some defect in his eyesight: for he adds that it was less distinct when he was
looking at it very intently than when he was making less effort.

140.   How the movement of a Planet begins.

Having related all those things concerning Comets, let us now return to
the Planets. Let us suppose that star N[153] is less solid, or capable of less
agitation {or force to continue its movement in a straight line} than the
globules of the second element near the circumference of our heaven, but
that it has somewhat more [solidity] than those close to {the center in which
is} the Sun. Given these conditions, we shall understand that as soon as N
has been carried away by the vortex of the Sun, it must continuously
descend toward the center, until it reaches {the point at which are [found]}
those heavenly globules which are equal to it in solidity, or in ability to
continue their movement along straight lines. When it is finally at that
point, it will neither move closer to nor farther from the Sun (unless driven
slightly this way or that by some other causes), but will constantly revolve
around the Sun among those heavenly globules {which are equal to it in

[153] See Plate VI.

force}, and will be a Planet. For if it descended closer to the Sun, it would there find itself surrounded by slightly smaller heavenly globules which it would exceed in force to recede from the center around which it revolves. These parts would also be more rapidly moved, which thus would increase its own agitation along with its force, causing it to ascend. If, on the other hand, it receded further from the Sun, it would encounter there heavenly globules which were somewhat less rapidly moved and would thus decrease its agitation, and which were slightly larger and would thus have the force to drive it back toward the Sun.

141.   The causes on which Planetary deviations depend: the first.

The first of these other causes which drive a Planet (thus suspended around the Sun) slightly in some direction, is that the space in which it is revolved along with all the matter of the heaven is not perfectly spherical; for the matter of the heaven must necessarily flow more slowly where this space is wider than where it is narrower, {thus permitting the Planet to move further away from the Sun}.

142.   The second.

The second cause is that the matter of the first element, by flowing toward the center of the first heaven from certain neighboring vortices, and thence back toward certain others, can displace in various ways both the globules of the second element and the Planet suspended among them.

143.   The third.

The third is that the pores in the body of this Planet may be more suited to admitting those grooved or other particles of the first element which come from certain areas of the heaven than to admitting the others. As a result, the entrances of these pores, which, as I indicated above,[154] are formed around the poles of the spots covering stars, turn toward those areas of the heaven rather than toward the others.

144.   The fourth.

Another cause is that there may have previously been some movements in the Planet which it still retains long afterward, even though the other

[154] Article 105.

causes oppose this. For we see that a spinning top acquires enough force, merely from the fact that a boy twirls it once, to continue subsequently to spin on its own for several minutes, and to rotate during that time several thousand times {around its axis}, even though it is very small and even though both the air which surrounds it and the earth on which it presses oppose its movement. Similarly, one can easily believe that if a Planet had been set in motion from the moment of its creation, that alone would be sufficient to allow it to continue its rotations from the beginning of the universe up to the present time without any significant decrease in speed,[155] because {the greater a body is, the longer it can retain the agitation which has been communicated to it in this way; and because} the five or six thousand years for which the universe has existed are a much shorter time compared to the size of a Planet than a minute is compared to the tiny bulk of a spinning top.

145.   The fifth.

Finally, the fifth of these causes is that the force to thus continue in its movement is much more stable and unchanging in a Planet than in the heavenly matter surrounding it, and it is also more stable in a large Planet than in a smaller one. For of course the force of this heavenly matter depends on the fact that its globules simultaneously unite in the same movement. Since they are separated from one another, in a few moments it can occur that sometimes more and sometimes fewer of them thus simultaneously unite. From this it follows that the Planet is never moved as rapidly as the [individual] globules surrounding it. For even though it may equal [that portion of] their movement by means of which it is transported along with them, the globules meanwhile have many other movements, since they are separated from one another. It also follows from this that when the motion of these heavenly globules is accelerated, or slowed, or turned aside; the motion of the Planet situated among them is not so greatly or so easily increased, slowed, or turned aside.

---

[155] This seems to be an approach on Descartes's part to the principle of conservation of angular momentum. Burman reports that, in conversation, Descartes explicitly stated that rotating bodies would rotate perpetually if not impeded by other surrounding bodies. See A. & T., V, 173. Cf. also Articles 48, and 150.

146.   Concerning the creation of all the Planets.

If all these things are {thoroughly} considered, nothing {which it has so far been possible to observe} concerning the phenomena of the Planets will occur which is not perfectly in accordance with the laws of nature which we set forth, and whose reason is not easily provided {and deduced} from what has already been said. For nothing prevents us from judging that that extremely vast space which now contains the vortex of the first heaven was formerly divided into fourteen or more vortices. And [we may judge that] these vortices were so arranged that the stars which they had at their centers gradually became covered by many spots, and that then some of these vortices were destroyed by others in the manner which has been described; some sooner and some later depending on their diverse situations. So that, since those three vortices which had at their centers {those bodies which we now call} the Sun, Jupiter, and Saturn were larger than the others; the stars in the centers of the four smaller vortices surrounding Jupiter descended toward Jupiter {to become the four little Planets which we now see there}. Then, there were also two others close to that of Saturn, and their stars descended toward it in a similar fashion (at least if it is true, {as seems to be the case}, that there are two {other lesser} Planets rotating around Saturn).[156] {And when the vortex containing it was destroyed, the Moon also descended toward the Earth}; and when the vortices which had Mercury, Venus, the Earth . . . ,[157] and Mars, at their centers were destroyed {by another larger one, in the center of which was the Sun}, they descended toward the Sun, {and there arranged themselves in the manner in which they now appear}. Finally, Jupiter and Saturn, together with the smaller stars joined to them, also assembled near the same Sun (which was much larger than they were), after their vortices had been destroyed. However, the Stars of any remaining vortices, if there were ever more than fourteen in that space, {having become more solid than Saturn}, were transformed into Comets.

---

[156] When viewed at an angle through an inadequate telescope, the rings of Saturn appear as two fuzzy globes on either side of the planet (they have the annoying habit of disappearing entirely when the rings happen to be edgewise to the Earth). They were first correctly identified as broad, flat rings by Huygens in 1655, thereby explaining their curious behavior.
[157] The Latin text adds "the Moon" here; the French text has altered this in order to make the sequence of the formation of the Earth-moon system parallel that of Jupiter and Saturn.

147.   Why certain Planets are further from the Sun than others, and
            that this does not depend solely on their size.

   Thus, when we now see the principal Planets, Mercury, Venus, the Earth,
Mars, Jupiter, and Saturn being transported around the Sun at different
distances, we shall judge that this occurs because {they are not all equally
solid, and that} those which are closer to the Sun are less solid than those
further away. And we have no reason to think it strange that Mars,
although smaller than the Earth, is further from the Sun, because size is not
the only factor which determines the solidity of bodies, so that Mars,
{though smaller}, can be more solid than the Earth.

148.   Why those closer to the Sun move more rapidly than the others;
            and yet the Sun's spots move very slowly.

   And, seeing that the Planets which are closer to the Sun are revolved
more rapidly than those which are further away, we shall think that this
occurs because the matter of the first element which forms the Sun, rotating
extremely rapidly {on its axis}, carries the parts of the heaven which are
close to it along with it more violently than it does those which are further
away. And, in spite of this, we shall not think it strange that the spots which
appear on its surface move more slowly than any Planet. (The spots take
about twenty-six days to complete their orbit, which is very short; while
Mercury takes less than three months to complete its orbit which is more
than sixty times as long. And Saturn completes in thirty years an orbit
which it would not finish in a hundred if it moved no more rapidly than the
spots, since its path is approximately two thousand times as long as theirs.)
For we shall think that this occurs because those particles of the third
element which are formed by the continuous dissolution of the spots have
accumulated around the Sun and these form a great mass of air or aether,
which reaches to the sphere of Mercury, or {perhaps} even further. And the
particles of which this aether is composed, having very irregular branching
figures, become attached to one another in such a way that they cannot be
moved individually like the globules of heavenly matter, but are all carried
along simultaneously by the Sun, and with them both the Sun spots and
also the part of the heaven near to Mercury. As a result, they do not
complete many more revolutions than does Mercury in the same space of
time, and therefore are not being moved as rapidly.

149.   Why the Moon revolves around the Earth.

Since the Moon revolves not only around the Sun but also around the Earth at the same time, we shall judge that this occurs either because the Moon moved toward the Earth before the Earth was transported around the Sun, in the same way that the Planets of Jupiter flowed toward Jupiter. Or, perhaps more correctly, we shall judge that this happens because the Moon has as much force of agitation as the Earth; thus, it must be situated in the same orbit[158] around the Sun: and, since its bulk is less and it has the same force of agitation, it must move more rapidly. For, if the Earth is situated near the Sun S,[159] on the circle NTZ and is transported along it from N through T toward Z, and if the Moon is being moved more rapidly [in the same direction] and arrives at the same circle; at whatever point on the circle NZ it first happens to be, it will shortly reach A, where, due to the nearness of the Earth {and the resistance of the air and the part of the Heaven which surrounds the Earth}, it will be prevented from continuing in a straight line and will turn its course toward B. I say toward B rather than toward D because in that way its course will deviate less from the straight line.[160] However, while it is thus proceeding from A toward B, all the heavenly matter contained in the space ABCD, which carries the Moon along, will be rotated around center T like a vortex; {which has rotated ever since}. This will also cause the Earth to rotate on its axis,[161] while at the same time all these things, {the Earth, the Moon, and this space of the heaven}, will be transported around center S along the circle NTZ.

[158] Literally: 'sphere'.
[159] See Plate XV.
[160] That is, the line tangent to NTZ at point A.
[161] It is not clear from the text whether the moon rotates the matter in the space, or whether its rotation carries the moon; possibly both are intended. In 1646, Clerselier sent Descartes a number of objections to the *Principles*, which had been made by Le Conte. Le Conte, objecting to Principle 149, stated that there seemed no reason why the moon would not simply continue in its orbit and strike the Earth; and certainly no reason to suppose that its direction of motion would then be *reversed*, supposing that it had passed over the Earth and reached C. Descartes does not seem to have replied to this; but Picot, who had seen the objections before they were sent to Descartes, wrote: "What prevents the Moon from approaching the Earth so closely as to touch it is the heavenly matter which gives the Moon so much agitation when it approaches A that it recedes from the Earth and forms its own vortex. The reason why it does not move toward Z, when it is at C, is that it is more easily moved inside that vortex [ABCD, presumably] than outside it, since the heavenly matter is more agitated there;...": A. & T., IV, 464–465. Though obviously confused, this may be the best reply one could make without invoking some sort of force of attraction.

150.   Why the Earth is rotated on its axis.

However, this may not be the only thing which causes the Earth to rotate on its axis. For, {since we are considering it as} if it had once been a bright star occupying the center of some vortex, we must think that it then rotated in this way, and that now the matter of the first element, which accumulated in its center, still has similar movements, and drives it.

151.   Why the Moon moves more rapidly than the Earth.

And there is no reason to think it strange that the Earth makes almost thirty rotations on its axis in the time it takes the Moon to complete the circle ABCD only once, because the circumference of this circle is approximately sixty times as long as the Earth's path,[162] so that the Moon still moves twice as rapidly as the Earth. Moreover, since both are transported by the same heavenly matter, which probably moves at least as rapidly in the vicinity of the Earth as in that of the Moon, I think that the only reason why the Moon moves more rapidly than the Earth is that it is smaller.

152.   Why, as nearly as possible, the same side of the Moon is always
              turned toward the Earth.

We should not marvel, either, at the fact that the same side of the Moon always faces the Earth, or certainly is at least never much turned away from it.[163] We shall easily judge that this occurs because the far side of the Moon is somewhat more solid, and therefore must complete a larger orbit as it revolves around the Earth; on the model of what was noted earlier concerning Comets. And certainly those innumerable inequalities, resembling mountains and valleys, which are observed on the near side with the aid of a telescope, seem to prove that that side is less solid {than its other side may be}. The cause of this smaller degree of solidity may be that the far side, which never comes within our view, receives light only directly from the Sun; while the near side also receives that which is reflected from the earth.

[162] The Earth's circumference is meant here; that is, the moon moves at twice the speed that an object on the surface of the Earth is rotated.
[163] For various reasons, a total of almost 59% of the moon's surface is visible during the course of a lunar month.

153.   Why the Moon travels more rapidly, and deviates less from its mean motion [when] in conjunction than in quadrature, and why its heaven is not spherical.

Neither shall we marvel at the fact that the Moon is seen to be somewhat more rapidly moved, and to deviate less from its course in any direction when it is full or new (that is, when it is near parts B or D of the Heaven); than when only half of it is visible, that is, when it is near A or C.[164] Since the heavenly globules in the space ABCD differ in size and motion from those which are below D near K and from those above B near L, but are similar to those near N and Z; they spread more freely toward A and C than toward B and D. From this it follows that the orbit ABCD is not a perfect circle, but closer to the figure of an ellipse;[165] and that the matter of the heaven is transported more slowly when in the regions near C and A than when near B and D. Therefore, the Moon, which is carried along by this matter of the heaven, {must also move more slowly and deviate more from its course near C and A and} must move closer to the Earth if it is waxing and further away if it is waning;[166] that is, further away when it happens to be near A or C, than when it is near B or D.

154.   Why the secondary Planets which are near Jupiter move so quickly; and why those near Saturn move so slowly, or not at all.

Furthermore, we shall not wonder that the {two} Planets said to be near Saturn revolve around it only very slowly, or perhaps not at all; while, on the contrary, the four which are near Jupiter rotate {rapidly} around it: each the more rapidly the closer it is to Jupiter. For we can believe that this diversity is caused by the fact that Jupiter, like the Sun and the Earth, turns on its axis; while Saturn, {the most elevated of the Planets}, always keeps

---

[164] See Plate XV. This phenomenon is known as the lunar variation and was first detected by Tycho Brahe; it is due to the effect of the sun's gravity on the Earth-moon system. Descartes does not seem concerned with any of the other variations in lunar motion, which were well known at the time.

[165] This is not, of course, a Keplerian ellipse, since the Earth is at the center rather than at a focus.

[166] In terms of the diagram, the moon will be waxing when it is moving from half (A) to full (B) and waning when it is moving from full to half (C).

the same side turned toward the center of the vortex containing it, as do the Moon and the Comets.[167]

155.   Why the poles of the Ecliptic and the Equator are very distant
           from each other.

Moreover, we shall not wonder that the axis around which the Earth rotates in the space of a day, is not situated perpendicularly to the plane of the ecliptic on which it revolves around the Sun in the space of a year, but deviates from the perpendicular by more than 23°; from which fact results the diversity of summer and winter on the earth. For the annual movement of the Earth on the ecliptic is mainly determined by the common course of all the heavenly matter revolving around the Sun, as is obvious from the fact that all the Planets follow this course {along the ecliptic} as nearly as possible. However, the direction of the Earth's axis around which its daily rotation occurs depends more on the [location of those] areas of the heaven from which the matter of the first element flows toward the Earth. For of course, since we are imagining all the space which is now occupied by the first heaven to have been formerly divided into fourteen or more vortices, in the centers of which were those stars which have now become Planets, we cannot think that the axes of all these stars were turned in the same direction; for this would not be in agreement with the laws of nature.[168] However, it is very likely that the matter of the first element which used to flow toward the Earth when it was a star, came from more or less the same areas of the firmament which its poles now face. And [it is likely that] when many layers of spots were gradually being formed on that star, the grooved particles of this matter of the first element bored many passages for themselves in these layers, and adjusted these to their size and figure so that the grooved particles coming from other areas of the firmament either cannot be admitted by these passages or can be admitted only with difficulty. Consequently, since those which bored passages suited to themselves through the globe of the Earth parallel to its axis still constantly flow through it; they cause its poles to be turned toward those areas of the heaven from which they come.

[167] This would seem to indicate that it is the Earth's rotation which rotates the vortex carrying the moon; but cf. Article 149.
[168] See Article 65.

156.   Why they are gradually moving closer to one another.

Meanwhile, however, because the Earth's two rotations, annual and daily, would be more conveniently accomplished if performed around parallel axes; the causes preventing this are gradually changing on both sides. Thus, with the passing of time, the obliquity of the Ecliptic to the Equator is decreasing.[169]

157.   The final and most general cause of all the inequalities observed in the movements of bodies in the universe.

Finally, we shall not wonder that all the Planets slightly deviate in every way, both longitudinally and latitudinally, from those perfectly circular motions which they are always attempting. For, inasmuch as all the bodies in the universe are contiguous and act on one another, {there being no possibility of any void}, the movement of each is affected by the movements of all the others and therefore varies in innumerable ways.[170] And I think that there is absolutely no phenomenon which is observed in the distant heavens, which has not been sufficiently explained here. It remains now for us to explain {in a similar way} those things which are observed near to us on the Earth.

[169] See note 35.
[170] This seems to be the first occurrence of the view that every heavenly body could affect the motions of all others.

PART IV

OF THE EARTH

PART IV

1.   That the false hypothesis which we have already used must be
       retained here, in order to explain the true natures of things.

Although, as I have already sufficiently warned, I do not wish it to be
believed that the bodies of this visible world were ever created in the
manner which was described above; I must however still retain the same
hypothesis, in order to explain the things which are seen on the Earth. So
that if, finally, as I hope to do, I clearly show that the causes of all natural
things can be understood by means of that hypothesis, though by no other;
it will thence be justly concluded that their nature is the same as if they had
indeed been formed in such a way, {although the world was not formed in
that way in the beginning, but was created directly by God}.

2.   How the Earth was created, according to this hypothesis.

And so let us imagine that this Earth which we inhabit was formerly {a
star} like the Sun, composed solely of the matter of the first element,
although it was much smaller than the Sun; and that it was situated in the
center of a vast vortex. But, since the grooved particles of this matter of the
first element, and other tiny particles of it which were not the smallest of all,
adhered to one another; they were thereby transformed into the matter of
the third element. At first, opaque spots were created from these parts on
the surface of the Earth, similar to those which we see being constantly
produced and destroyed around the Sun. Then, [let us imagine] that the
particles of the third element which remained from the continuous
disintegration of these spots, having spread over the nearby heaven,
eventually formed there a great accumulation of air, or aether. Finally,
after this aether had become very extensive, the denser spots which had
been formed around the Earth entirely covered and darkened it. And since
these spots could no longer be destroyed, and since they were perhaps very
numerous and rested upon one another, and since the force of the vortex
containing the Earth was meanwhile diminishing; the Earth ultimately

descended, along with the spots and all the air enveloping it, into another larger vortex, in the center of which is the Sun.[1]

3.   The division of the Earth into three {diverse} regions; and a
            description of the first of these.

If we consider the Earth in the state in which it must have been a short time before thus descending toward the Sun, we shall be able to distinguish in it three very different regions. It seems that the first and innermost of these, {here designated as} I,[2] must contain only the matter of the first element, which is moved there in the same way as that which is in the Sun. And it does not differ in nature from that which is in the Sun except for the fact that it is perhaps not quite as pure, since it cannot purify itself, as does that of the Sun, by continually expelling the matter of the spots. And because of this, I would be inclined to believe that the whole of the space I must now be filled almost exclusively by the matter of the third element, {formed by the least subtle parts of the first, as they became attached to one another}, if it did not seem to follow that if that were the case, the Earth would be so solid that it could not remain as close to the Sun as it does.[3] {In addition, one can imagine various reasons why it is impossible for space I to contain anything other than the purest matter of the first element; for perhaps the parts of this matter which are the most inclined to adhere to one another are prevented, by the body of its second region, from entering space I. Perhaps, too, when this matter is confined within this space, its movement has so much force that it not only prevents any of its parts from remaining united, but also gradually detaches some particles from the body which surrounds it}.

4.   A description of the second.

The intermediate region M is entirely occupied by a very opaque and dense body: for, inasmuch as this body was formed from the tiniest particles (i.e., those which previously belonged to the first element) joined to one another; no passages seem to have been left in it unless they are so

[1] The French text has the order of events somewhat reversed here, stating that the fourteen planetary vortices were within the vortex of the sun before their stars darkened. This is not the order of events described in both the Latin and French versions of Articles 118 and 119 of Part III, however.
[2] See Plate XVI, Fig. i.
[3] See Part III, Article 147.

small that they can only admit those grooved particles described above and the remaining matter of the first element. And experience shows that this is true of the Sun spots, which, since they are of the same nature as body M (except that they are much thinner and more rarefied) nevertheless impede the penetration of light. They could scarcely do this if their passages were sufficiently large to admit the globules of the second element. For, seeing that these passages were in the beginning formed in fluid or soft matter {which consists of particles which are very small and pliable}, the passages would undoubtedly also be sufficiently straight and smooth not to impede the action of light.

5.   A description of the third.

But these two inner regions of the Earth concern us very little, because no one has ever reached them alive. There remains only the third region, from which, as we shall successively demonstrate, all the bodies which are here found around us can originate. Now, however, we are supposing that this third region is still nothing other than a great accumulation of the particles of the third element, having much of the heavenly matter {of the second} among them. And the innermost nature of these particles can be known from the way in which they were created; {and thus we shall also be able to arrive at a perfect knowledge of all the bodies which must be composed of them}.

6.   That the particles of the third element which are in this third region must be quite large.

Of course, because they resulted from the disintegration of the spots {which at one time formed on the Earth when it was still similar to the Sun and} which were composed of the tiniest scrapings of the first element joined to one another: each of these particles must be composed of many scrapings of that kind. And they must be sufficiently large to have withstood the impulse[4] of the globules of the second element which moved around them; because any whatsoever which could not do so would have been transformed again into particles of the first or the second element.

[4] Latin: `impetus`.

7.   That these particles can be changed by the first and second
           element.

But although these particles as a whole resist the globules of the second
element, nevertheless, because the {very small 'and flexible} individual
scrapings of which they are composed yield to those globules; some of the
parts of these particles can always be affected by an encounter with the
globules.

8.   That they are larger than the globules of the second element,
           but less solid and less agitated.

And since these scrapings of the first element have diverse figures, it was
not possible for a great number of them to be joined together so closely that
the particles of the third element thus formed did not contain many very
narrow pores, which only the most subtle matter of the first element could
penetrate. As a result, although these particles are much larger than the
heavenly globules, they cannot however be as solid or capable of as much
agitation. The fact that they have extremely irregular figures, less suited to
movement than are the spherical figures of these globules, also contributes
to this. For, inasmuch as the scrapings of which they are composed are
joined together in innumerable diverse ways; it follows that these particles
must differ very much from one another in size, solidity, and figure; and
that practically all of their shapes must be extremely irregular.

9.   That, from the beginning, these particles pressed upon one
           another around the Earth.

And here it must be noted that while the Earth was situated in its own
individual vortex like the fixed stars and had not yet descended toward the
Sun, those particles of the third element which enveloped the Earth,
although separate from one another, were not however scattered randomly
in all directions through its heaven. Rather, having all accumulated around
sphere M, they pressed upon one another. This is because they were being
driven toward center I by the globules of the second element; which
{formed a vortex around this Earth and were more massive and thus} had a
greater force of agitation than the particles of the third, and were striving to
recede from that center.[5]

[5] This phenomenon will become the basis of Descartes's explanation of weight; see Article 23.

10.   That various interstices left between these particles have been
             filled by the matter of the first and second element.

It must also be noted that, although they thus pressed upon one another,
they were not however joined so perfectly as not to leave very many
interstices around themselves. These were not only occupied by the matter
of the first element, but also by the globules of the second: for this had to
follow from the fact that these particles had extremely irregular and diverse
figures, and had joined together in a disorderly manner.

11.   That in the beginning, the parts of the second element were
             smaller, the closer they were to the center of the Earth.

Moreover, we must note that among those {parts of the second element
which were situated in these interstices}, the lowest, {in relation to the
Earth}, were somewhat smaller than those higher up, for the same reason
that it was stated above[6] that those around the Sun become gradually
smaller, the closer they are to the Sun. And all these {parts of the second
element, which were in the Earth's highest region}, were no larger than
those which are now around the Sun below the sphere of Mercury; but were
perhaps smaller, because the Sun is larger than the Earth has ever been.
From which it follows that they were also smaller than those which now are
here {in this same region of the Earth}, because the latter, being further
from the Sun than those below the sphere of Mercury, must consequently
be larger.[7]

12.   And that they had narrower passages between [the particles of
             the third element].

And it must be noted that, {as the terrestrial parts of this highest region
were formed}, those globules retained for themselves, between the particles
of the third element, paths adapted to the size {of these smallest particles of

[6] See Part III, Article 85.
[7] The claim here is that those globules which were close to the Earth when it was in its own
vortex were smaller than those which are now close to the sun. However, once the Earth
entered the sun's vortex, its surrounding globules were driven toward the sun to a location
appropriate to their size and were replaced by larger globules of a size appropriate to the
Earth's present distance from the sun. The French text of this article is somewhat clearer, and
has been preferred.

the second element}; so that other slightly larger globules could not so easily pass through them.

13.   That the bulkier parts {of this third region} were not always
                   lower than the less solid[8] ones.

Finally, we must notice that, at that time, it frequently happened that the larger and more solid of these parts of the third element were situated above others which were smaller and less solid; because they had only one uniform movement around the Earth's axis, and easily adhered to one another on account of the irregularity of their figures. Thus, although each one was being driven toward the center {of the Earth}, and although the larger and more solid each was, the greater the force with which it was driven by the globules of the second element; nevertheless, the more solid could not always free themselves from the less solid in order to descend below them. Thus, those particles frequently retained approximately the same order in which they had been formed; {so that those coming from the spots which were the last to be destroyed were the lowest}.

14.   Concerning the first formation of diverse bodies in the third
                   region of the Earth.

Afterwards however, when the globe of the Earth, divided into these three regions, descended toward the Sun (the vortex in which it was formerly situated having of course been consumed), no great change in its innermost and intermediate regions could have resulted from this descent. However, the exterior region must have been altered into first two, then three, presently four, and subsequently very many other diverse bodies.

15.   Concerning the actions by means of which these bodies were
                   created; and first, concerning the general movement of the
                   heavenly globules.

I shall shortly explain the creation of these bodies; but, before I can undertake that, the three or four most important actions on which this depends must be considered. The first is the movement, considered in general, of the heavenly globules. The second is weight. The third is light.

[8] It is unclear what sense of 'solid' is involved here; cf. Articles 3 and 8, and Part III, Article 121.

And the fourth is heat. By the general movement of the heavenly globules, I understand their continuous agitation; which is so great that it not only suffices to carry them around the Sun in annual movement, and around the Earth in daily movement, but also to move them meanwhile in very many other ways. And because, in whatever direction they have begun to be moved, they subsequently continue in that direction along lines which deviate as little as possible from the straight; these heavenly globules, having mingled with the parts of the third element forming all the bodies of the third region of the Earth, produce various effects in the latter: the three most important of which I shall note here.

16.   Concerning the first effect of this first action, which is to make bodies transparent.

The first effect is that the heavenly globules render transparent all those terrestrial bodies which are liquid and composed of particles of the third element so rarefied {and thus sufficiently far apart so} that these globules are carried around them in all directions. For since these globules are constantly being moved from all directions through the passages in these bodies, and have the force to change the situation of their particles; they easily make in those bodies paths for themselves which are straight or equivalent to straight, and thereby suited to the transference of the action of light. And thus we know with certainty from experience that there are no liquids on Earth which are pure, and composed of rarefied particles, but which are not transparent. For as for quicksilver, its particles are too bulky to admit the globules of the second element around them on all sides, {but only those of the first}. As for ink, milk, blood, and such things, they are not pure liquids but are interspersed with many small grains of hard bodies. And as concerns hard bodies, it can be observed that all those are transparent which were transparent while they were being formed and were still liquid, and whose parts retain the same situation in which they were placed by the globules of the heavenly matter moving around them when they did not yet adhere to one another. On the other hand, all those bodies whose particles were simultaneously joined and entwined by some external force, which did not conform to the movement of the heavenly globules mingling with them, are opaque. For although many passages remain in these bodies through which the heavenly globules constantly travel from all directions; these passages are interrupted and closed off in various places, and cannot be suited to the transmission of the action of light, which is only carried along straight lines or lines equivalent to straight.

17.   How a solid and hard body can have enough passages to
             transmit rays of light.

And in order that it may be understood here how hard bodies {such as
glass or crystal} can have a sufficient number of passages to transmit rays of
light coming from any direction, let us imagine that some apples, or any
other fairly large, smooth globes, are enclosed in a mesh bag which
squeezes them closely together in such a way that these apples, adhering to
one another, form, as it were, a single body. In whatever direction this body
is turned, there will be contained in it passages through which small lead
balls, which have been thrown onto it, will easily descend by the force of
their weight toward the center of the earth, along lines equivalent to
straight. {And even if we accumulate so many such pellets in this hard body
that all the passages in it are filled with them; when the pellets higher up
press upon those beneath them, the action of their weight will pass in a
straight line to the lowest ones}. Thus this bag of apples will simulate a
transparent body which is solid and hard. For it is not necessary for the
heavenly globules to find passages in the terrestrial bodies through which
they transmit rays of light which are straighter or more numerous than
those through which the small lead balls descend between these apples.

18.   Concerning the second effect of this first action: which
             separates some bodies from others, and purifies liquids.

The second effect is that, when the particles of two or more terrestrial
bodies, especially liquids, are confusedly joined together, the heavenly
globules tend to separate some of them from one another, and thus to
divide them into different bodies. However, they also tend to mix certain
others more completely, and arrange them in such a way that each droplet
of the liquid composed of these different particles will become exactly
similar to all the other droplets of that same liquid. For when the heavenly
globules are being moved through the passages of terrestrial liquid bodies,
they constantly drive some of the particles of the third element which they
encounter out of their place, until they have so disposed and arranged these
among the others that they do not oppose the movements [of the globules]
more than those others do; or, when the particles cannot be so disposed,
until they have separated them from the rest. Thus we see that the lees of
unfermented wine are driven out, not only to the surface and to the bottom
(which could be attributed to lightness and weight), but also toward the

sides of the vessel. And after the wine has been purified {by the action of this subtle matter}, it is transparent, although still composed of diverse particles; and it does not appear denser or thicker at the bottom than at the top. And the same is to be thought about the other pure liquids.

19.   Concerning the third effect, which makes drops of liquids spherical.

The third effect of the heavenly globules is that, as I have already explained in the [*Discourses on*] *Meteorology*,[9] they make drops of water or other fluids, which are suspended either in the air or in another fluid different from them, spherical. For since these heavenly globules have very different paths in a drop of water than in the surrounding air, and always, as far as is in their power, are moved along lines as straight as possible; it is evident that, upon encountering a drop of water, those globules which are in the air are less impeded from continuing their movements along lines as straight as possible if that drop is perfectly spherical than if it has any other figure whatever. For if there is some part on the surface of this drop which protrudes beyond the spherical figure, the heavenly globules travelling through the air will strike against that part with greater force than against the others, and therefore will thrust it toward the center of the drop. And if some part of the surface of the drop is nearer to the center than the rest, the heavenly globules contained in the drop itself will drive that part away from the center with greater force;[10] and thus all the heavenly globules contribute to making the drop spherical. And, since the angle which a circular line makes with a straight one which is tangent to it is smaller than any rectilinear angle, and is everywhere equal in no curved line except the circular: it is certain that a straight line can never be more evenly curved, and less curved at every one of its points, than when it changes into a circular one. {Thus, movements which are prevented from being straight by a cause which acts on them equally at all points, must be circular when they occur along a single line, and spherical when they occur toward all sides of some surface}.

[9] See Discourse V.
[10] The French text explains that this occurs because the globules within the drop are more impeded by an indented portion of its surface than by a spherical portion.

20.   The explanation of the second action, which is called weight.

The force of weight does not differ much from this third [effect of the first] action of the heavenly globules. For, in the same way as these globules, solely by means of the movement by which they are carried along equally in every direction, thrust all the particles of each drop equally toward its center, and thus make the drop itself spherical: so, through the same movement, when they have been prevented, by encounter with the whole bulk of the earth, from being transported along straight lines; these globules drive all its parts toward the middle: and in this consists the weight of terrestrial bodies.

21.   That the parts of Earth, considered individually, are not heavy, but light.

In order that the nature of weight may be perfectly understood, it must be noted that if all the spaces around the Earth which are not occupied by the matter of the Earth itself were empty, that is, if they contained nothing except a body which in no way either helped or hindered the movement of other bodies (for only thus can the word 'empty' be understood), and if, meanwhile, the earth were rotating on its axis by its own movement in the space of twenty-four hours; all those parts of the Earth which were not very firmly attached to one another would fly off in all directions toward the heaven. In the same way it can be seen that when a spinning-top is turning, if sand is thrown onto it, this sand immediately is thrown back from it and is dispersed in all directions. And {if that were the case}, the Earth should not be said to be heavy, but, on the contrary, would have to be said to be light.

22.   In what the lightness of the heavenly matter consists.

However, since there is no such emptiness, and since the Earth is not carried along by its own movement, but is moved by the heavenly matter which surrounds it and which penetrates all its pores,[11] the Earth has the

---

[11] The French text of this article explicitly states that the movement in question is the Earth's daily rotation rather than its yearly revolution. This would not seem to be in accord with Article 144 of Part III, nor with the Latin of this article. Further, the view that weight results from a circulation of heavenly matter which rotates the Earth would seem to imply that objects should fall toward the Earth's axis rather than toward its center; cf. Article 23.

mode of a body which is at rest. The matter of the heaven, in so far as it all unites in that movement by which it carries the Earth along, has no force either of weight or of lightness; but in so far as its parts have more agitation than they employ in moving the Earth, and since they are always prevented, by encounter with the Earth, from continuing their movement along straight lines, {they use this excess agitation both to turn more rapidly than the Earth . . . and to perform diverse other movements in other directions, and} they always recede from {the center of} the Earth as much as they can; and their lightness {in relation to the parts of the Earth} consists in this.

23.   How all parts of the Earth are driven downward by this heavenly matter, and thus become heavy.

Next, it must be noted that the force which the individual parts of the heavenly matter have to recede from {the center of} the Earth cannot produce its effect unless, while those parts are ascending, they press down and drive below themselves some terrestrial parts into whose places they rise. For, seeing that all the spaces which are around the Earth are occupied either by particles of terrestrial bodies or by the heavenly matter; and seeing that all the globules of this heavenly matter have an equal propensity to move away from the Earth: individually they have no force to drive other similar globules from their place. However, since such a propensity is not as great in the particles of terrestrial bodies; whenever the heavenly globules have some of these terrestrial particles above them, the former must bring this force of theirs to bear upon the latter in every way. Thus, the weight of each terrestrial body is not, strictly speaking, produced by all the heavenly matter flowing around it, but rather only by that portion of the heavenly matter which immediately ascends into the place of the descending body, and which, therefore, is exactly equal to it in size. For example, if B[12] is a terrestrial body suspended in mid-air which is united with more particles of the third element than a quantity of air equal to it, and which therefore has fewer or narrower pores in which the heavenly matter is contained; it is evident that if this body B descends toward I, a quantity of air equal to it must ascend into its place. And because this quantity of air contains more of the heavenly matter {which is striving to recede from the Earth's center} than there is in B, it is also evident that there must be, in this

---

[12] See Plate XVI, Fig. i.

quantity of air, the force to drive B down {and thus to give it the quality which we call weight}.

24.   How much weight there is in each body.

And in order that this calculation may be correctly made, it must be observed that in the pores of this body B, there is also some of the heavenly matter, and that this is opposed to {and has as much force as} an equal quantity of similar heavenly matter contained in the mass of air; and this renders that quantity of heavenly matter in body B useless {so that only the excess [of heavenly matter in the air] should be counted}. Similarly, there are some terrestrial parts in the mass of air, which are opposed to an equal number of the other terrestrial parts of body B, and have no effect on them. However, when these things have been subtracted on both sides, what remains of the matter of the heaven in this mass of air acts against what remains of the terrestrial parts in body B; and B's weight consists in this alone.

25.   That weight does not correspond to the quantity of matter in each body.

And, in order that nothing may be omitted, it must also be noticed that, by 'heavenly matter', I mean here not only the globules of the second element, but also the matter of the first mingled with them. And those terrestrial particles which are following the course of the heavenly matter (and thus are more rapidly moved than the rest) are also to be placed in this category; such are those which form the air. Besides, the matter of the first element, other things being equal, has greater force to drive terrestrial bodies downward than do the globules of the second, because it has more agitation; and for a similar reason, these globules have greater force than {a similar quantity of} the terrestrial particles of air which they move with them. As a result, it cannot easily be estimated from weight alone how much terrestrial matter is contained in each body. And it may be that although a piece of gold may weigh twenty times as much as a quantity of water equal to it [in volume], yet it may not contain four or five times as much terrestrial matter: not only because an equal amount must be subtracted from both because of the air in which they are being weighed,[13]

---

[13] According to Archimedes' principle, a body immersed in a fluid loses an amount of weight equal to the weight of the volume of that fluid which it displaces.

but also because in water, as in all other fluid bodies, because of the motion of their particles, there is lightness in comparison with hard bodies.

26.   Why bodies do not gravitate [when] in their natural places.

It must also be observed that in all motion there is a circle of bodies which are moved together, as has already been shown, and that no body is borne down by its own weight unless another lighter body, equal to it in size, is borne upward at the same moment of time. As a result, in a vessel which is extremely deep and wide, the lower drops of water, or of another liquid, are not pressed upon by the higher ones; nor are individual parts of the bottom of the vessel pressed upon, except by as many drops as rest upon them perpendicularly. For example,[14] in vessel ABC, drop of water 1 is not pressed upon by the others 2, 3, 4, situated above it, because if these were carried downward, other drops 5, 6, 7, or similar, would have to ascend into their place; and since these drops are equally heavy, they {hold the former in equilibrium and} prevent their descent. However, these drops 1, 2, 3, 4, having united their forces, press upon part B of the bottom;[15] because, if they cause it to descend, they also will descend, and parts of air 8, 9, which are lighter than they are, will ascend. But no more drops than {are in the cylinder} 1, 2, 3, 4, or others equivalent, press upon this same part B of the vessel; because at the same moment of time at which this part B can descend, no other drops can follow it. And from this it is extremely easy to explain innumerable observations concerning the weight of bodies, or rather, if it is permissible to speak thus, their gravity; observations which seem astonishing to those who philosophize poorly.

27.   That weight drives bodies down toward the center of the Earth.

Finally, it must be noted that, although the particles of heavenly matter are agitated at the same time by many diverse movements, yet all of their actions harmonize and, as it were, counterbalance one another in such a way that, due solely to their encounter with the bulk of the earth which resists their movements, they strive to move away equally in all directions from its vicinity, as if from its center; unless by chance some exterior cause

[14] See Plate XVI, Fig. ii.
[15] How these drops all press on area B yet not on each other is most unclear; Descartes may have thought that their force would be united only if there were a hole at B, but that is not what is said.

introduces diversity into this matter. Now, several such causes can be imagined; but I have not yet been able to ascertain[16] whether their effect is sufficiently great to be perceived by the senses.

28.   Concerning the third action, which is light; and how it agitates
                 the particles of air.

The force of light, insofar as it spreads to all parts of the heaven from the Sun and stars, has already been sufficiently explained above: it only remains for us to note here that those of its rays which have come from the Sun agitate the Earth's particles in diverse ways. For in fact, although the force of light, considered in itself, is nothing other than a certain pressure which occurs along straight lines drawn from the Sun to the Earth: yet because this pressure is not applied equally to all the particles of the third element which form the highest region of the earth, but now to some and now to others, and even sometimes to one extremity of a particle and sometimes to the other: it can easily be understood how various movements in such particles are caused by that pressure. For example,[17] if AB is one of the particles of the third element forming the highest region of the earth and is resting upon another particle C, and if between AB and the Sun, there lie many other particles, like D, E, F: these intervening particles now prevent the Sun's rays G, G, from pressing upon extremity B, but not, however, from pressing upon [extremity] A. Thus extremity A will be pushed down, and the other, B, raised up. And because these particles are constantly changing situation, a little while afterwards, they will oppose the Sun's rays striving to move toward A, but not, however, the other rays heading toward B, and thus extremity A will be raised up again, and B will be pushed down. This same thing occurs in all the particles of the earth which the Sun's rays reach; and therefore they are all agitated by the Sun's light.

29.   Concerning the fourth [action], which is heat: what it is, and
                 how it remains when light has been removed.

However, this agitation of the Earth's particles, whether it originates from light or from any other cause, is called heat; especially when it is

[16] The French text reads, "... but I have not yet been able to perform any experiment which would indicate whether ..." here.
[17] See Plate XVI, Fig. iii.

greater than usual and affects the senses; for the name 'heat' relates to the sense of touch. And it must be noted that each one of the terrestrial particles which has been thus agitated continues in its movement, in accordance with the laws of nature, until it is halted by some other cause, {for heat consists solely in the movement of the particles of bodies}; and therefore, heat produced by light always remains for some time after the light has been extinguished.

30.   Why heat penetrates further than light.

In addition, it must be noted that the terrestrial particles, which have thus been driven by the Sun's rays, agitate other nearby particles to which these rays do not penetrate; and that these agitate still others; and so on. And since an entire half of the Earth is always illuminated by the Sun; so many particles of that kind are simultaneously agitated that, although light halts on the Earth's opaque outer surface, the heat generated by that light must nevertheless penetrate to the innermost parts of the {second or} intermediate region of the Earth.

31.   Why it rarefies practically all bodies, {and why it also condenses some}.

Finally, it must be noted that these terrestrial particles, when more than usually agitated by heat, cannot generally be contained in such a small space as when they are at rest or less agitated; because they have irregular figures, which occupy less space when they are at rest and joined in some particular way, than when they are disunited by continual motion. As a result, heat rarefies practically all terrestrial bodies, but some more and some less, depending on their various situations and on the shape of the particles of which they are composed. {So that there are also bodies which heat condenses, because their parts are better arranged and move closer together when those parts are more agitated, as was said of ice and snow in the *Discourses on Meteorology*}.[18]

---

[18] See Discourse VI, where Descartes offers this as an explanation of the expansion of water when it freezes.

32.   How the highest region of the Earth was first divided into two
             different bodies.

Once these various actions {which can cause changes in the order of the
Earth's particles} have been noticed, let us again consider the Earth as just
now approaching the vicinity of the Sun, and as having its highest region
composed of particles of the third element which are not firmly joined
together and with which are mingled heavenly globules which are
somewhat smaller than those which are found in that part of the heaven
through which the Earth is passing, or even in that to which it is coming.
We shall then easily understand that these smaller globules will give up
their places to the somewhat larger ones which surround the Earth, and
that these somewhat larger ones, rushing violently into these places, {which
are too narrow to admit them easily}, strike against many particles of the
third element, especially against the bulkier ones whose weight contributes
to this effect, and drive these below the rest. This causes those bulkier
particles which have been driven below the rest, and which have irregular
and varied figures, to be more closely fastened together and to interrupt the
motion of the heavenly globules more than those higher up. As a result, the
highest region of the Earth, such as is here shown at A,[19] will be divided
into two very different bodies, such as are shown at B and C: the higher one,
B, being rare, fluid, and transparent, while the lower one, C, is {in
comparison} quite dense, hard, and opaque.

33.   The division of the terrestrial particles into three principal
             species.

Then, from the fact that we judge that body C was separated from body
B solely because its parts, having been pressed down by the heavenly
globules, adhered to one another; we shall also understand that still
another body, such as D, must have subsequently been created between
these two. For indeed, the figures of the particles of the third element, of
which bodies B and C are composed, are extremely varied, as was noted
above, and we can here distinguish these particles into three principal
kinds. Some of these particles are assuredly made up of various arm-like
parts, which extend this way and that, as do the branches of trees and other
things of that sort; and these particles are the principal ones which, having

[19] Articles 32 through 40 refer to Plate XVII, the order of formation of the Earth occurring
counter-clockwise from A.

been driven down by the matter of the heaven, begin to adhere to one another and to form body C. Others are more solid, and have {more compact} figures, not indeed all spheres or cubes, but rather as angular as any crushed stones. And if these are fairly large, they descend beneath the others by the force of their weight {and unite with body C}; if, however, they are fairly small, they remain mingled with the particles of the first kind {in body B}, and occupy the interstices left by these. Finally, the last kind are rather long and branchless, like staffs: and these also intermingle with the particles of the first kind,[20] when they find sufficiently large intervals between them, but are not easily joined to them; {and thus can easily be moved relative to them}.

34.   How a third body was formed between the first two.

Once these things have been noticed, it is in accordance with reason for us to believe that when the branching particles of body C first began to be entwined with one another, they had most of the fairly long particles {of the third kind} intermingled with them. Subsequently, however, when the branching particles were squeezed more and more {by the action of the matter of the Heaven} and became more closely joined, the long particles ascended above the branching ones toward D, where they accumulated together in a body very different from B and C. In the same way, we see that when one treads upon the earth in marshy places, water is squeezed out of it and afterwards covers its surface. Neither is it doubtful that, meanwhile, many other particles {of both the third and second kind} fell down from body B and augmented the bulk of the two lower bodies C and D.

35.   That this body [D] is entirely composed of particles of one kind.

Although in the beginning these fairly long particles and also others which were solid, like rude bits or fragments of rock, were mingled with the branching ones; it must however be noted that the more solid particles did not as easily ascend above the branching ones as did the fairly long ones. Or, if some did ascend, they subsequently more easily descended beneath the branching ones again. Since the fairly long particles, other things being equal, have more surface-area in proportion to their bulk; they therefore are more easily driven out by the matter of the heaven flowing through the pores of body C: and {because they are long}, after they have reached D,

---

[20] The French text states that these wand-like particles are found in both body B and body C.

lying crosswise upon the surface of body C, they do not easily encounter pores through which they could return into C, {as did the parts of the second kind}.

36   That there are only two species of particles of this kind in it.

And thus many fairly long particles of the third element accumulated at D. At first they were neither perfectly equal to one another nor perfectly similar; however, they had in common the fact that they could not adhere easily either to one another or to other particles of the third element, and that they were moved by the heavenly matter flowing around them. As a result of this latter property, they withdrew from body C and accumulated together at D. And since the heavenly matter constantly flowed around them there, and caused them to be agitated by various movements, and caused some [particles] to migrate into the places of others, they must have become smooth and polished and as nearly as possible equal to one another with the passing of time, and have been reduced to only two species. Namely, those which were sufficiently thin that they could be bent solely by that impetus by which they were being driven by the heavenly matter; and these twisted around other slightly bulkier particles, which could not be bent by that impetus, and which carried the former along with them. And these two species of particles, the flexible and the inflexible, having thus been joined, continued in their movement more easily than either could have done alone. As a result, it happened that both have remained in body D, {and were not reduced to a single species}. Further, those which, in the beginning, could be bent around others were rendered more and more flexible by being bent {by the matter of the Heaven}, and became like eels or short slender cords with the passing of time. Since the others were never bent, they gradually lost any flexibility they might once have had and remained rigid like spears.

37.   How the lowest body C was divided into several others.

Further, it must be thought that body D began to be separated from the other two bodies B and C before these two had been completely formed; that is, before C was so hard that its particles could no longer be more closely joined or driven lower by the motion of the heavenly matter, and before the particles of body B were all so arranged as to offer, around themselves on all sides, easy and even paths {along straight lines} to this matter of the heaven: and that therefore many particles of the third element were afterwards still driven out of body B toward C. And these particles, if they were more solid than those which had congregated at D; descending below these, united with body C, and, depending on the diverse nature of their figures, either remained on its surface or penetrated below [that surface]: and thus this one body C was divided into several others: and also perhaps became entirely fluid in one of its regions, because the particles which had congregated there had figures which prevented them from adhering to one another. But all things cannot be explained here.[21]

38.   Concerning the formation of another fourth body above the third.

Also, when particles which were less solid than those which formed body D fell down out of body B {by the action of the matter of the Heaven}, they adhered to the surface of D. And because most of these [particles] were branching, they gradually became joined to one another and formed a hard body E, very different from the fluid bodies B and D. And this body E was at first extremely thin, like a rind or shell covering the surface of body D. With time, it became thicker, because of new particles from body B joining themselves to it, and also particles from D (which were not exactly similar to the other particles of D, and thus were driven out of it by the motion of the heavenly globules), as I shall soon explain. And because of the various actions of light and heat, {which caused these particles in E to rise and descend}, these particles were differently arranged in those parts of the earth where it was day or summer, than in those where it was night or winter. Thus, whatever approached this body [E] on one day or during one summer was in some way distinguished from whatever approached the same body on the following day or during the following summer; and thus E was formed of various quasi rinds or shells stretched over one another.

39.   Concerning the accretion of this fourth body and the purification of the third.

And a long time was certainly not necessary for the highest region A of the Earth to be divided into two bodies B and C; nor indeed for many fairly long particles to accumulate around D; nor, finally for the first interior

---

[21] The final sentence of the French text reads, "But it is impossible to explain everything."

shell of body E to be formed. But only in the space of many years could the particles of body D have been reduced to the two types described a short while ago, and all the shells of body E be formed. For, in the beginning, there was no reason why some of the particles which were flowing toward D should not be slightly bulkier and longer than others, or why they should be absolutely smooth and polished; they could instead have had a certain roughness, although not so much as to cause them to become connected to the branching particles. And they could also have been either rough or smooth along their length, and bulkier at one end than at the other. Since they did not adhere to one another, however, and since, therefore, the heavenly matter constantly flowing around them had the force to move them; most of them gradually became smooth and polished, by rubbing against one another[22]; and [became] equal to one another in size and cylindrical along their length, because they were passing through the same paths and succeeding one another in places which could neither admit larger particles nor be entirely filled by smaller ones. But also very many, which could not be reduced to the common pattern of the others, were gradually driven out of this body D by the motion of the heavenly globules; and quite a few of these particles {which were more solid than the others in D} certainly united with body C; but most, {which were less solid}, ascended toward E and B, and augmented body E by increasing its matter.

40.   How this third body was diminished in bulk and left some space between itself and the fourth body.

And in fact, during the day and during the summer, when the Sun was (by the force of light and heat) rarefying one half of body D, all the matter of this half could not be contained between the two neighboring bodies C and E; nor could it drive these hard neighboring bodies from their places. Therefore, a great many of the particles of this matter ascended through the pores of body E toward B; but when this rarefaction ceased at night and during the winter, they descended again on account of their weight. However, there were many causes why all the particles of the third element which were thus leaving body D could not subsequently return into it. For

---

[22] The French text omits the phrase "by rubbing against one another"; probably because it is not clear from the Latin which of these effects are due to the rubbing and which are due to the size of the passageways.

they had greater force when leaving than when returning; because the force of dilation produced by heat is greater than the force of weight. And therefore, many particles made a path for themselves through the narrow pores of body E, along which to ascend; but subsequently finding no path along which to return, they remained on E's surface. And also several [ascending particles], having become trapped in these pores and not being strong enough to ascend further, closed those pores to those which were about to descend. Further, any which were thinner than the rest and differed sufficiently from their smooth and polished figure, were driven out of body D solely by the motion of the heavenly globules, and therefore were the first to ascend toward E and B. Upon encountering the particles of these bodies, they not infrequently changed their figures and either adhered to the particles of these bodies or at least ceased to be suited to returning toward D. From which it must have followed, after many days and years, that a great part of this body D had been consumed and that no particles were found in it except those of the two types previously described. Further, body E must have become quite dense and thick, because almost all the particles which had receded from D, having become stuck in E's pores, had made it denser; or else, having been changed by encountering and adhering to the particles of body B, had fallen down again toward E, and thus had increased its thickness. Finally, a fairly large space F[23] was left between D and E. And this space could not be filled by matter other than that of which body B was composed; the slenderest particles of which could, of course, easily pass through the pores of body E into the places which were being vacated by other slightly larger particles leaving D.

41.   How numerous fissures were made in the fourth body.

Thus, although body E was heavier and denser than F (and perhaps also than D), because of its hardness, it remained suspended for a time above D and F like a vault. But it must be noted that when E first began to be formed, it had very many pores hollowed out in it which were sufficiently large to admit the particles of body D. For, seeing that E was then leaning upon D's surface, it could not help offering passage to those particles which were moved each day by the force of heat and ascended toward B during the day, and descended {by their weight} during the night, and which thus

[23] See Plate XVIII, Fig. i.

constantly filled these pores. However, after body D had decreased in bulk, its particles no longer occupied all the pores of body E; and other smaller particles, coming from B, took their places. And since these smaller particles did not sufficiently fill these pores of body E, and since no void is possible in nature; the heavenly matter, which is the only thing which can fill all the narrow interstices which are found around the particles of terrestrial bodies, rushed into these pores, changed their figures, and created the impetus to thrust some of them apart in such a way that the other neighboring pores were made narrower. As a result, it easily happened that, when certain parts of body E had been separated from one another, fissures were created which became larger and larger with the passing of time. This occurred for exactly the same reason that many cracks appear in the earth when it is dried out by the Sun in the summer, and that the earth opens up more and more, the longer the drought lasts.

42.   How it was broken into various parts.

However, since there were many such cracks in body E, and since these were constantly increasing in size, finally its parts adhered to one another so insufficiently that it could no longer be supported between F and B like a vault; and therefore E, entirely broken, fell by its own weight onto the surface of body C. And since this surface was not large enough to receive all the adjacent fragments of E in the situation which they previously had had, some of these fragments must have been inclined to one side and resting upon one another. For example, if the portion of body E which is illustrated had its principal fissures situated at points 1, 2, 3, 4, 5, 6, 7; and if two fragments 23 and 67 began to fall down a little before the rest, and the extremities 2, 3, 5, and 6 of four other fragments began to fall before the opposite extremities 1, 4, and V; and similarly, extremity 5 of fragment 45 fell down somewhat sooner than extremity V of fragment V6: there is no doubt that these [fragments] must now be arranged on the surface of body C in the way in which they are here depicted.[24] Thus, fragments 23 and 67 are most closely joined to body C, while the other four are lying on their sides and resting upon one another, etc.

[24] See Plate XVIII, Fig. ii.

43.　　　How the third body in part ascended above the fourth, and in
　　　　 part remained below it.

Nor is there any doubt that body D, which is fluid and less heavy than the
fragments of body E, must occupy all the lower cavities left beneath these
fragments, and also their fissures and pores; but, in addition, because D
cannot be entirely contained in these, there is also no doubt that it ascends
above the lower of these fragments, for example, 23 and 67; {and by the
same means must have formed passages for itself in order to enter or leave
from the underneath of some to the top of others}.

44.　　　That, as a result, mountains, plains, oceans, etc., were created
　　　　 on the surface of the Earth.

Now, if we understand that air is here[25] taken to be bodies B and F; a
certain very thick {and very heavy} interior crust of the earth, from which
metals originate, to be C; water to be D; and the exterior earth, composed
of stones, clay, sand, and mud, to be E: we shall also easily understand the
water covering fragments 23 and 67 to be the oceans; other fragments
which are only gently inclined and not covered by any water, like 89 and
VX, to be the level ground of plains; and the fragments raised up higher,
like 12 and 94V, to be the mountains. And, finally, we shall notice that
when these fragments fell down in this way by the force of their own weight,
their extremities were violently dashed against one another and were
divided into many other smaller fragments which formed cliffs on certain
shores of the oceans, as for example at 1, and numerous mountain peaks,
sometimes very lofty, as at 4, sometimes lower as at 9 and V, and also
shelves of rock in the ocean, as at 3 and 6.

45.　　　What the nature of air is.

And the innermost natures of all these things {such as air, minerals, and
all other bodies on the Earth} are implicit in[26] what has already been said.
First, we know from the preceding that air must be nothing other than an
accumulation of particles of the third element, so thin and separated from
one another that they obey all the movements of the heavenly globules; and
that therefore air is a very rare, fluid, and transparent body and can be

[25] See Plate XVIII, Fig. ii.
[26] Literally, "... can be drawn out (*eruo*) from what has been said."

formed of particles of any figure at all. Indeed, if its particles {could adhere
to one another and} had not been completely disunited from one another,
they would have long ago adhered to body E. Since they are disunited, each
one is moved independently of its neighboring ones; and thus occupies that
whole small spherical space which it requires for its circular movement
around its own center, and drives all neighboring particles out of that
space. This is why it does not matter {for this effect} what figures the
particles of air may have.

46.   Why it is easily rarefied and made dense.

Air, however, is easily made dense by cold and rarefied by heat: for, since
almost all of its particles are flexible, like soft feathers or thin cords; the
more rapidly they are driven, the more widely they extend themselves, and
therefore require a larger sphere of space for their movement. And it must
be noted, in accordance with what has been said, that by heat, simply the
acceleration of movement in these particles must be meant, and by cold, its
diminution.[27]

47.   Concerning its forcible compression in certain machines.

Finally, air which is forcibly compressed in some vessel has the {same}
force to spring back {as was employed to compress it} and to extend itself
immediately into a wider space. On the basis of this, machines are created
which, by means of air alone, hurl water upward, as {very elevated} springs
do;[28] others hurl shafts with great impetus, as bows do.[29] The cause of this
is that, when air has thus been compressed, each of its parts does not have
that small spherical space which it needs for its movement to itself, but

[27] The French text states, "... it [heat] should increase their agitation and cold should diminish
it." in apparent contrast to the Latin and to Article 29.
[28] The reference here is apparently to the upward surge of the water which has been piped from
a higher elevation.
[29] In order to correct what the translator considered a mistake, the French text substitutes,
"And small guns [are built] which, charged only with air, drive balls or arrows almost as
vigorously as if they were charged with powder," for this sentence. Very powerful compressed
air muskets were used as early as 1530. However, a stone-throwing catapult in which the bow
arms were powered by compressed air was at least designed by Ctesibius of Alexandria in the
third century B.C. Ctesibius also designed a compressed air pump for use in fountains. The
pump is described in Vitruvius' *Architecture* (Venice, 1567), which was in the library at La
Flèche while Descartes was a student there.

other neighboring particles enter into that space. Since the same heat or the same agitation of these particles is meanwhile maintained by the heavenly globules constantly flowing around the particles of air, the latter strike one another with their extremities, and attempt to drive one another out of their place, and thus together produce the force to occupy a greater space. [30]

48.   Concerning the nature of water, and why it is easily transformed; sometimes into air, and sometimes into ice.

As for water, I have already shown that only two species of {long and smooth} particles are found in it, some of which are flexible, while the others are inflexible: and that if these [two types of particles] are separated from one another, some form salt and the others sweet water. [31] And, because I have already fully explained, in the [Discourses on] Meteorology, [32] all the properties, both of salt and of sweet water, which are deduced from this one foundation; there is no need for me to write more about them here. But I only wish it to be noted how perfectly all these things fit together and how it follows from such a creation of water that there must also be such a proportion between the size of its particles and the size of those of air, and likewise between these particles and the force with which the globules of the second element move them; that when these globules drive them a little less than usual, they change water into ice, and the particles of {the vapors found in} air into water; however, when they drive them slightly more vigorously, the slenderest particles of water, namely those which are flexible, are transformed into air. [33]

49.   Concerning the ebb and flow of the ocean.

I have also explained, in the [Discourses on] Meteorology, [34] the causes of the winds, by which the ocean is agitated in various irregular ways. But there remains another regular movement of the ocean, by which, twice a day in each place, it is raised up and driven down, and meanwhile always

[30] This article and the one preceding constitute a first step toward the kinetic theory of gases.
[31] Descartes is referring here to the Discourses on Meteorology rather than to a previous section of the Principles.
[32] Discourses III and V.
[33] The French text treats air as being a mixture of pure air and water vapor; but the Latin clearly claims that an actual transformation, resulting simply from a change in agitation, occurs.
[34] Discourse IV.

flows from the East toward the West. In order to explain the cause of this movement, let us visualize that small vortex of the heaven which has the Earth as its center, and which is carried along in a larger vortex around the Sun with the Earth and the Moon. And let ABCD be that small vortex;[35] EFGH, the Earth;1234, the surface of the ocean, which, for the sake of greater clarity, we are supposing completely covers the Earth; and 5678, the surface of the air encompassing the ocean. Now if there were no Moon in this vortex, point T, which is the center of the Earth, would be at point M, which is the center of the vortex; but when the Moon ☾ is situated near B, this center T must be between M and D. Since the heavenly matter of this vortex is moved somewhat more rapidly than the Moon or the Earth, [both of] which it carries along with it; if point T were not somewhat more distant from B than from D, the presence of the Moon would impede this heavenly matter from being able to flow as freely between B and T as between T and D. And since the location of the Earth in this vortex is determined only by the equality of the forces of the heavenly matter flowing around it, it is evident that the Earth must therefore approach D to some extent. And in the same way, when the Moon is at C, the center of the Earth will have to be between M and A; and thus the Earth will always recede slightly from the Moon. Further, in this way, not only is the space through which the heavenly matter flows between B and T made narrower by the Moon at B, but so is the space through which the heavenly matter flows between T and D. It follows that this heavenly matter flows more rapidly in those spaces and therefore presses more upon the surface of the air at 6 and 8, and upon the surface of the water at 2 and 4, than it would if the Moon were not on diameter BD of the vortex. And since the bodies of air and water are fluid and easily obey this pressure, these bodies must be less deep above parts F and H of the Earth than if the Moon were not on diameter BD; and, on the contrary, [these bodies] must be deeper at G and E, so that the surfaces of the water 1, 3, and of the air 5, 7, swell there.

50.   Why water ascends in 6 1/5 hours, and descends in 6 1/5 hours.

Now {because the Earth rotates once every twenty-four hours}, that part which is now at F (below the region of point B and where the ocean is as shallow as possible) will be at G after six hours (below the region of point C

---

[35] See Plate XIX, noting that the small circle located at B represents the moon, and that the dotted ellipse around M represents the path of the Earth's center during one synodic month.

where the ocean is deepest); and after six more hours at H, below the region of point D; and so on. More precisely (since the Moon meanwhile slightly advances from B toward C, because it describes circle ABCD in one month), that part of the Earth which is now at F (below the region of the body of the Moon) will, after approximately six hours and twelve minutes, be beyond point G, [lying on] that diameter of vortex ABCD which intersects the diameter of the vortex on which the Moon then lies at right angles; and then the water will be deepest there. And after another six hours and twelve minutes, it will be beyond point H, in the place where the water will be as shallow as possible, etc. From this, it may be clearly understood that the water of the ocean must ebb and flow in one and the same place every twelve hours and twenty-four minutes.[36]

51.   Why the ocean's tides are greater when the Moon is full or new.

It must be noted that this vortex ABCD is not exactly round, but that the diameter on which the full or new Moon is situated is shorter than the diameter which intersects it at right angles, as has been shown above.[37] It follows thereby that the ebb and flow of the ocean must be greater when the Moon is full or new than during the intervening times.[38]

52.   Why they are greatest at the equinoxes.

It must also be noted that the Moon is always on a plane near to the Ecliptic, whereas the Earth is rotated by its daily movement along the plane of the equator; these two planes intersect each other at the equinoxes, while at the solstices they are very distant from each other. It follows from this

---

[36] Descartes's explanation of the tides rather neatly accounts for two somewhat unrelated effects. First, there are two high and two low tides every twenty-four hours and fifty minutes, which is the average interval between two passages of the moon across the meridian line, the point at which it is highest above the horizon. Second, the tidal bulge of the ocean arrives in some places as much as six hours after the moon has reached its highest point in the sky, thus low tide may occur when the moon is overhead. This effect is due to the rotation of the Earth and to irregularities in the ocean bed and shoreline. The lag between the overhead passage of the moon and the arrival of the tidal bulge attracted by the moon varies greatly from place to place, but is seldom more than six hours; cf. Article 55.

[37] See Part III, Article 153.

[38] In fact, this is because when the moon is new or full the sun and moon both lie on roughly the same line through the Earth; thus, the attractive power of the sun is added to that of the moon.

that the greatest tides of the ocean must occur around the beginning of
Spring and Autumn, {when the Moon acts most directly against the
Earth}.[39]

53.   Why air and water always flow from the East to the West.

Furthermore, it must be noted that while the Earth is being rotated from
E via F toward G, or from West to East, the swelling of water 412, and
similarly the swelling of air 856, which are now resting upon part E of the
Earth, gradually thereby travel toward other more Westerly parts of it.
Thus, after six hours {and twelve minutes}, they will rest upon part H of the
Earth, and after twelve hours {and twenty-four minutes}, upon part G; and
the same thing must also be understood of the swellings of water and air 234
and 678. As a result, water and air are carried along by a continuous flow
from the Eastern parts of the Earth to its Western parts.[40]

54.   Why, in the same latitude, regions which have the ocean on the
                 East are more temperate than others.

This flow, although admittedly not extremely rapid, is however distinctly
perceived from the fact that long sea voyages are much slower and more
difficult in Easterly directions than in Westerly ones; and also from the fact
that the water always flows toward the West in certain straits of the ocean.
Furthermore, other things being equal, those regions which have the ocean
on the East, like Brazil, do not feel the heat of the Sun as much as those like
Guinea, which have long tracts of land on the East and the ocean on the
West; since the air coming from the ocean is colder than that coming from
land.

---

[39] The reasoning here seems obscure, to say the least. On Descartes's view, the height of tides
varies inversely with the distance between the Earth and the moon. There appears to be a
confusion here between the angular distance from the moon to the celestial equator and the
linear distance from the moon to the Earth's equator. His intention may be to argue that the
heavenly matter rotates most rapidly at the celestial equator and thus would be most impeded
when the moon lies on the equator; cf. Article 49. Near the time of the equinoxes, the location
of the sun in the sky is closest to and most directly opposite the new and full moon.
[40] The effect of the Earth's rotation on winds and currents is due to the Coriolis force, which
deflects moving objects (except those moving directly to the east or west) to the right in the
northern hemisphere and to the left in the southern. As a result, the prevailing winds in the
northern temperate zone are from northeast to southwest; cf. Article 54.

55.   Why there is no ebb or flow in lakes and ponds; and why it
           occurs on various shores at various hours.

Finally, it must be noted that the ocean does not in fact cover the whole
Earth, as we assumed a little earlier; but because the Ocean[41] extends
around the Earth's entire periphery, as far as the general movement of the
Ocean's waters is concerned, it must be understood as if the Ocean did
envelop the whole Earth. However, lakes and ponds which are cut off from
the Ocean undergo no movements of this kind: because their surfaces are
not so large as to be pressed upon much more in one area than in another by
the heavenly matter which is impeded by the Moon's presence. And
because of the inequalities of the gulfs and curving shorelines by which the
Ocean is encompassed, the increases and decreases of its waters reach
diverse shores at diverse hours, giving rise to innumerable differences in its
tides; {although the waters in the middle of the Ocean rise and fall
regularly, in the way I have described}.[42]

56.   How the particular causes [of tides] {can be explained and} must
           be investigated on individual shores.

We shall be able to deduce the particular causes of all these diversities
from what has been said, if we consider that, when the Moon is new or full,
the waters of the Ocean, at places distant from the shores and near the
Ecliptic and the Equator, are [their] highest at the sixth hour, both of the
morning and of the evening.[43] These waters accordingly flow toward the
shores; but at the twelfth hour [these waters] are lowest, and accordingly
flow back from the shores toward these places. And depending on whether
the shores are nearby or far off, and whether the waters head for them along
paths which are more straight or oblique, wide or narrow, deep or shallow;
the waters will be carried to the shores more quickly or more slowly, and in
greater or smaller quantity. And also, on account of the extremely varied
and uneven contours of the shores, it often happens that the waters heading
toward one shore meet those which are coming from another, so that their

---

[41] The term used at this point in the Latin is 'Oceanus', which refers to the collective total of the
Earth's oceans. Previously, when referring to the ocean, the term 'mare' (uncapitalized) has
been used. The French reflects this change as well.

[42] There would, of course, have been no way of establishing this in Descartes's time.

[43] At the equinoxes, the sun would rise and set at the sixth hour; the new or full moon, being
close to or opposite the sun, would then set or rise at roughly those times and be above or
beneath the observer at noon and midnight.

course is changed in diverse ways. And, finally, various winds, some of
which are usual in certain places, drive these waters in diverse ways. For I
think that nothing is anywhere observed concerning the flow and ebb of the
ocean, such that its cause is not included in these few remarks.

57.   Concerning the nature of the Earth's interior.

With regard to the interior earth C,[44] it can be noted that this is
composed of particles of any figure whatever, which are so bulky that the
globules of the second element do not carry these particles along with them
by their ordinary movement, but simply.render them heavy by driving them
downward and agitate them somewhat by passing through the numerous
pores which are found between them. The matter of the first element, filling
the narrowest of these pores, also agitates these particles; and so do the
terrestrial particles of the higher bodies D and E, since particles of D and E
often descend into the widest of C's pores and carry off with them some of
the bulkiest particles of C. And certainly it is credible that C's upper surface
is composed of branching particles which are very firmly attached to one
another; inasmuch as, while this body was being formed, they were the first
to restrain and weaken the impetus of the heavenly globules travelling {in
straight lines} through bodies B and D. However, there are nonetheless very
many interstices among C's particles which are sufficiently wide to permit
particles of salt, sweet water, and also other angular and branching
particles (which have fallen down out of body E), to pass through them.

58.   Concerning the nature of quicksilver.

However, below this [upper] surface, the parts of body C adhere less
closely to one another; and perhaps, at a certain distance below this
surface, there may have accumulated many particles which have figures so
smooth, rod-like, and polished that, although they rest upon one another
because of their weight, and, unlike the parts of water, do not allow the
globules of the second element to flow around them on all sides; they are,
however, {in no way attached to one another, but are} easily agitated by the
smaller of those globules which find some spaces between, and especially so
by the matter of the first element, which fills all the narrowest corners left

---

[44] See Plate XVIII.

there. And therefore, these particles form a fluid which is very heavy and not transparent, such as quicksilver is.

59.   Concerning the inequality of the heat penetrating the interior of the Earth.

Furthermore, just as we see that those spots which are generated around the Sun each day have extremely irregular and varied figures; thus too it must be judged that the intermediate region M of the Earth, which is composed of matter similar to these spots, is not everywhere equally dense. And therefore, it offers passage to a greater quantity of the matter of the first element in certain places than in others. And this matter of the first element, {coming from the center of the Earth and} passing through body C, agitates its parts more vigorously in certain places than in others. Similarly, the heat produced by the Sun's rays which penetrates (as has been stated) to the innermost Earth, does not act uniformly against this body C; because this heat is more easily communicated to C through the fragments of body E than through the water D. Also, the height of mountains causes certain parts of the Earth which {face South and} are turned toward the Sun to grow much hotter than those which are turned away from it. Finally, those parts near the Equator grow hot to a different degree than do those near the poles; and this heat sometimes varies on account of the alternation of day and night, and, especially, of summer and winter.

60.   Concerning the action of this heat.

As a result, the particles of this interior earth C are always being moved somewhat, sometimes more and sometimes less. And this is not only true of particles which are not joined to neighboring ones, like those of quicksilver, or salt, or sweet water, or others similar, {which descended from E and are} contained in the larger pores of body C. It is also true of those particles which are the hardest of all, and which adhere to one another as firmly as possible. Not that the latter become entirely separated from one another {by the action of heat or} by this movement; rather, as we see that branches of trees which have been shaken by the winds are agitated, and the interstices between them are made now larger, now smaller, although these trees are not torn from their roots: so it must be thought that the bulky and branching particles of body C are so connected and entwined that they are

not usually completely disunited from one another by the force of heat, but
are only struck against one another to some small extent, and the pores left
around them become now larger, now smaller. And since C's particles are
{much} harder than those which have fallen from the higher bodies D and E
into its pores, they easily break and diminish the particles of D and E by this
movement, and thereby reduce them to two types of figures, which must
now be considered.

61.　　　　Concerning the acrid and acid juices from which are formed
　　　　　vitriol,[45] alum,[46] {and other such minerals}.

Now of course, particles whose matter is a little more solid, such as those
of salt, which are caught and bruised in these pores, are transformed from
rod-like and rigid into flat and flexible: just as a cylindrical rod of white-hot
iron can be flattened out into a fairly long blade by repeated blows from a
hammer. And since these particles are meanwhile agitated by the force of
heat and are moving slowly this way and that through these pores; after
being struck and rubbed by the hard walls of the pores, they become
sharpened like swords, and thus are transformed into certain acrid, acid,
corroding juices. These juices, subsequently uniting with metallic matter,
form vitriol; with stony matter, alum; and form many other substances in
the same way; {depending on whether they mingle, as they congeal, with
metals, stones, or other materials}.

62.　　　　Concerning {the formation of} the oily matter of bitumen,[47]
　　　　　sulphur, etc.

However, the softer particles (such as the very numerous ones which fell
down from the exterior earth E and also those of sweet water) become so
thin, after having been thoroughly crushed in these pores, that they are torn
to pieces by the movement of the matter of the first element, and are divided
into many extremely tiny and very flexible branching particles. These
branching particles {are carried toward body E and} may adhere to other

[45] 'Vitriol' refers to any of the metallic sulfate salts of sulphuric acid, which was known as 'oil
of vitriol'.
[46] Alum is aluminum sulphate or aluminum potassium sulphate.
[47] 'Bitumen' normally refers to asphalt or similar naturally occurring substances.

terrestrial particles, and thus form sulphur, bitumen, and all the other fatty or oily substances which are found in mines.

63.   Concerning the elements of Chemists; and how metals ascend into mines.

And thus we have here three things which {are closely related to and} can be taken to be the three customary elements of Chemists: salt, sulphur, and Mercury. That is, one may take the acrid juice to be their salt, the softest small branches of oily matter to be their sulphur, and quicksilver itself to be their Mercury. And it can be believed that all metals reach us only because acrid juices, flowing through the pores of body C, separate certain of its particles from these pores. Then, after these particles have become enveloped by and covered with oily matter, they are easily carried upward by quicksilver which has been rarefied by heat; and they form various metals according to their diverse magnitudes and figures. I would perhaps have described these individually here, if I had previously had the opportunity to perform the various experiments which are required for a certain knowledge of these metals.

64.   Concerning the exterior of the Earth and the origin of springs.

Let us now consider the exterior Earth E; which has certain of its fragments concealed beneath the ocean, while others are extended into plains, and still others raised up into mountains. And first let us note how easy it is to understand, with regard to such an exterior Earth, how springs and rivers originate; and how, despite the fact that these always flow into the ocean, their water never fails, nor does the ocean increase or grow sweet. For in fact, since there are great cavities full of water below the plains and mountains, there is no doubt that many vapors, i.e., particles of water separated from one another by the action of heat and rapidly moved, penetrate daily to the exterior surface of plains and to the highest peaks of mountains, Indeed, we see that many vapors of this kind are carried further, up to the clouds; and they surely must ascend more easily through the earth's pores, supported as they are by its particles, than through the air, whose fluid and mobile particles cannot thus support them. However, after these vapors have thus ascended, {and cannot rise any higher because their agitation decreases}, they become sluggish and lose heat. Once they have lost the form of vapor, they are transformed again into water, which cannot descend through the same pores through which vapor ascends, because these pores are too narrow. However, the water finds somewhat wider paths in the interstices of the crusts or shells of which the whole exterior earth is composed; and these paths lead the water obliquely along the declivity of valleys and plains. And where these subterranean paths of the waters come to an end on the surface of a mountain, a valley, or a plain, springs gush out; and when many of the resulting streams have joined together, they form rivers, and flow into the ocean via the more sloping parts of the surface of the exterior earth.

65.   Why the ocean is not increased by the fact that rivers flow into it.

However, although many waters are constantly flowing from the mountains toward the ocean, the cavities from which these waters ascend can never on that account be exhausted; nor can the ocean be increased. For this exterior earth could not have been created in the way described a bit earlier, that is, from fragments of body E falling {unevenly} onto the surface of body C, without the water D having retained for itself many very open pores beneath these fragments. Through these pores, a quantity of water equal to that which leaves the mountains always returns from the ocean toward the bases of the mountains. And thus, water circulates in the veins of the earth and in rivers in the same way as the blood of animals circulates in their veins and arteries.

66.   Why springs are not saline, and why the ocean does not become sweet.

And even though the ocean is saline, only particles of sweet water ascend into {most} springs; because of course these particles are slender and flexible, whereas the rigid and hard particles of salt cannot be easily transformed into vapors, and are not able to pass through the oblique pores of the earth in any way, {unless these pores are wider than usual}. And although this sweet water is continuously being returned to the ocean via the rivers, the ocean does not on that account become sweet, because an equal quantity of salt {which was left there when the vapors rose into the mountains} always remains in it.

67.   Why the water in certain wells[48] is saline.

Even so, we shall not be greatly astonished if much salt is perhaps found in certain wells very distant from the ocean. For since there are many gaping cracks in the earth, it can easily happen that salt water which has not been purified reaches these wells: either because the surface of the ocean is on the same level as the bottom of these wells, {in which case they usually participate in the tides}; or else because, where the paths are sufficiently wide, particles of salt are easily carried upward by the particles of sweet water, along the slopes of the hard body. This can be observed in a vessel, the mouth of which flares outward to some extent, such as in vessel ABC:[49] for while salt water is being evaporated in this vessel, {salt rises along the sides, and} all its edges usually become covered with a crust of salt.

68.   And also why salt is mined from certain mountains.

And from this it can also be understood how great masses of salt have formed like rocks in certain mountains. For in fact, since sea water ascends there, and the flexible particles of sweet water proceed further; the salt alone has remained and filled the cavities which happened to be there.

69.   Concerning niter[50] and other salts different from sea salt.

Yet sometimes, too, particles of salt spread through some fairly narrow pores of the earth, and there, losing something of their figure and quantity {and thus the form of common salt}, are changed into niter or sal-ammoniac[51] or similar {salts}. Indeed, very many fairly long and rigid unbranching particles of earth have had the forms of niter and other salts from their beginning. For these forms do not require anything other than that the particles [of salts] should be fairly long and neither flexible nor branching; and they form various species of salt, according as they vary in other respects.

[48] The French text uses the term *'fontaines'* both for the 'wells' of this article and the 'springs' of Article 66. the Latin is unambiguous, however, using two different terms.
[49] See Plate XVI, Fig. ii.
[50] The Latin term used here is *'nitrum'*, which can refer to a number of mineral alkalies. The French text translates this term as 'saltpeter', which is niter (potassium nitrate). The use of *'nitrum'* in Article 109, *et seq.* indicates that the interpretation of the French is correct.
[51] Sal-ammoniac is ammonium chloride.

70   Concerning vapors, spirits,[52] and exhalations rising from the interior earth to the exterior.

In addition to the vapors which are drawn out of waters concealed beneath the earth, many acrid spirits, oily exhalations, and vapors of quicksilver, ascend from the interior earth to the exterior; transporting with them particles of other metals. And all substances which are mined are formed by their mingling in various ways. By acrid spirits, I understand those particles of acrid juices, and also of sal-volatile,[53] which are separated from one another and moving so rapidly that the force by which they continue their motion in all directions overcomes their weight. By exhalations, however, I understand the extremely slender {and flexible} branching particles of oily matter, which have also been thus moved. For in fact, in waters, other juices, and oils, the particles move only very slowly; whereas in vapors, spirits, and exhalations they move very swiftly.

71.   How their various minglings create various kinds of stones and other minerals.

And indeed spirits fly in this way with greater force, and more easily spread through certain narrow pores of the earth, and adhere more firmly when trapped in these pores; therefore, they form harder bodies than do exhalations or vapors. And, since the diversity among these three things is extremely great because of the diversity of the particles of which they are composed; many kinds of stones and of other non-transparent minerals are created from them. This occurs when they are imprisoned in the narrow pores of the earth and adhere to and thoroughly mingle with the earth's particles. And whenever [unmingled] spirits thicken into juices in the cracks and cavities of the earth and then lose their most fluid and slippery particles, so that the remaining particles gradually adhere to one another; many kinds of transparent minerals and gems are thereby also created from them.

72.   How metals from the interior earth reach the exterior, and how cinnabar[54] is formed.

In the same way, vapors of quicksilver, by creeping through the small cracks and fairly large pores of the earth, leave particles of other metals

[52] Spirits are volatile substances, usually extracted by distillation.
[53] Sal-volatile is ammonium carbonate, although the term can also refer to a solution of ammonium carbonate in alcohol. The French text has, "...and volatile salts," here.
[54] Cinnabar is mercuric sulphide; it is the principal ore from which mercury is extracted.

which were mingled with them in these [cracks and pores]; and thereby the earth is impregnated with gold, silver, lead[55] and other metals. And these vapors, because of their extraordinary sliminess, then either penetrate further, or flow back down. However, they sometimes remain there when the pores through which they could return are blocked by sulphurous exhalations. And under these circumstances, the particles of quicksilver themselves become coated with a sort of extremely tiny fuzz of these exhalations, and form cinnabar. Furthermore, spirits and exhalations also transport several metals, such as copper, iron, and antimony, from the interior earth to the exterior.

73.   Why metals are not found in all areas of the earth.

And it must be noted that these metals ascend almost exclusively from those parts of the interior earth to which fragments of the exterior earth are immediately joined. As, for example, in this figure,[56] from 5 toward V; because metals cannot be transported by waters. As a result, metals are not found in all places indiscriminately.

74.   Why they are principally found at the bases of mountains facing South and East.

Furthermore, it must be noted that it is usual for these metals to be carried up toward the bases of mountains through the veins of the earth (as here toward V) and to accumulate mainly there; because in that particular place there are more cracks in the earth than elsewhere. And it is also usual for these metals to be accumulated more in those parts of mountains which are turned either toward the noonday Sun or toward the East than in the others; because heat, the force of which carries metals upward, is greater there. And, for that reason, it is also usual for metals to be sought by miners mainly in such places.

75.   That all mines are in the exterior earth; and that it is never possible to reach the interior earth by mining.

In addition, it must not be thought that it is ever possible to reach the interior earth by aby perseverance in mining: both because the exterior

[55] Mercury dissolves, or, more properly, forms an amalgam with, these and other metals.
[56] See Plate XVIII, Fig. ii.

earth is too thick, in comparison with human strength; and especially because of the intermediate waters, which would gush forth with greater impetus, the deeper the place in which their veins were first opened; and which would drown all miners.

76.   Concerning sulphur, bitumen, clay, and oil.

The sienderest particles of exhalations, such as those described a little while ago, form nothing except pure air when alone; but they are easily joined to the slenderer particles of spirits, and thus transform those smooth and slippery particles into branching ones. These branching particles, when mingled with acrid juices and certain metallic particles, constitute sulphur. And when they are mingled with particles of the earth and are also laden with various acrid juices, these branching particles form {earths which can burn, like} bitumen, {naphtha, etc.}; and when united only with particles of the earth, they form clay. And finally, {if they accumulate} alone, they are transformed into oil when their movement decreases to such an extent that they rest completely upon one another.

77.   How earthquakes occur.

But when these {exhalations} are too rapidly agitated to be thus transformed into oil, if by chance they flow in great quantity into the cracks and cavities of the earth, they there form dense and thick fumes, not unlike those which are given off by a recently extinguished tallow candle. {And just as these fumes are easily ignited if one moves the flame of another candle near them}; if by chance any spark of fire is struck in these cavities, these fumes are immediately ignited and become suddenly rarefied. Thus, they shake all the walls of their prison with great force, especially if many spirits are mingled with them: and thus earthquakes are produced.

78.   Why fire erupts from certain mountains.

It sometimes happens that while these movements are occurring, and after a part of the earth has been thrown asunder and opened, a flame erupts from the peaks of mountains toward the heaven. And these eruptions occur there, rather than in lower places: both because there are more cavities under mountains, and because those great fragments making up the exterior earth rest upon one another and offer an easier exit to the flame there than in any other places. And although this opening in the earth may be closed as soon as the flame has thus erupted from it, it can happen that a quantity of sulphur or bitumen which is sufficiently great to create an enduring conflagration has been driven out from the inmost parts of the mountain toward its summit. Moreover, new fumes, which have accumulated in the same cavities and have been ignited, easily erupt through that same opening subsequently. Consequently, several mountains are notorious for frequent conflagrations of this kind, including Etna in Sicily, Vesuvius in Campania, Hecla in Iceland, etc.

79.   Why many shocks usually occur in an earthquake: which, under these circumstances, lasts for several hours or days.

Finally, an earthquake sometimes lasts for several hours or days; because usually there is not just one continuous cavity in which the dense and flammable fumes are being collected, but many diverse cavities separated by earth filled with much sulphur or bitumen. So that when an exhalation has been ignited in some of these cavities and thus shaken the earth once, some delay intervenes before the flame can travel to the other cavities through the pores which are completely filled with sulphur.

80.   Concerning the nature of fire, and the differences between it and air.

However, it remains here for me to say how a blaze can be kindled in these cavities, and also to explain the nature of fire. When terrestrial particles, of whatever size or figure they may be, are individually and separately following the movement of the first element, they have the form of fire; just as they have the form of air when they are flying to and fro among the globules of the second element and share its agitation. Thus the first and principal difference between air and fire is that the particles of the latter are much more rapidly agitated than those of the former. For it has already been adequately shown above that the movement of the matter of the first element is much more rapid than that of the second. But there is also another very great difference. For while it is true that the bulkier particles of the third element, such as those which compose the vapors of quicksilver, can take on the form of air; they are not however necessary to its conservation. On the contrary, air is purer and less subject to corruption when it is composed of only the tiniest particles; for the bulkier particles fall downward because of their weight and of their own accord abandon the form of air unless they are agitated by continuous heat. Fire, on the other hand, cannot exist without being fed and renewed by rather bulky particles of terrestrial bodies.

81.   How fire is first kindled.

Now, the globules of the second element occupy all those interstices around the Earth which are sufficiently large to admit them, and they all rest upon one another in such a way that it is impossible to move some without the others (except perhaps circularly around their own axes). Thus, although the matter of the first element fills all the small corners left by these globules and is moved as rapidly as possible there; if it does not have more space than is contained in these corners, it cannot have sufficient force to carry along with it the terrestrial particles which are all supported by one another and by the globules of the second element, or, accordingly, to generate fire. In order for fire to be started somewhere, the heavenly globules must be forcibly expelled from the interstices of several terrestrial particles, which then become disunited from one another and are floating only in the matter of the first element. Thus, they are carried along by the very rapid movement of the latter, and are driven in all directions.

82.   How fire is maintained.

Moreover, in order that this fire may be maintained, these terrestrial particles must be sufficiently bulky, solid, and suited to motion, so that, after they have thus been driven by the matter of the first element, they have the force to repulse the heavenly globules from the place in which the fire is, and which they are attempting to re-enter. And only thus can the heavenly globules be prevented from again occupying the interstices left by the first element, and, by weakening the force of the latter in this way,[57] extinguishing the fire.

83.   Why fire requires fuel.

Moreover, the terrestrial particles which strike against these globules cannot be prevented by them from proceeding further {into the air}; and

[57] The French text simply has, "...and only thus can the globules be prevented from extinguishing the fire..." here; probably because the Latin is somewhat unclear.

thereby leaving that place in which the first element is exerting its force, losing the form of fire, and disappearing into smoke. For this reason, no fire would last long if some of these terrestrial particles, striking against some body bulkier than air, did not at the same time detach other sufficiently solid particles from that body. These take the place of the first ones, and, being carried rapidly along by the matter of the first element, continually generate a new fire. {And this must happen as rapidly as the fiery particles are being transformed into smoke. It is also necessary for the parts of this body to be sufficiently numerous and large to have the power to repulse those parts of the second element which strive to stifle the fire: and air alone cannot achieve this, which is why air is insufficient to maintain fire}.

84.   How fire is struck from flints.

In order that these things may be more precisely understood, let us first consider the various ways in which fire is generated; secondly, all the things which are required to maintain it; and finally, what its effects are. Nothing is more usual than for fire to be struck from flints; and I judge that this occurs because flints are fairly hard and rigid, and at the same time fairly friable.[58] Since they are hard and rigid {(i.e., such that if their parts are even slightly bent, they tend to resume their original shape)}, if they are struck by another hard body {such as a steel}, the spaces which lie between many of their particles, and which are normally occupied by the globules of the second element, become narrower than usual. Therefore these globules, having been forced out of the spaces, leave nothing around the particles of flint except the pure matter of the first element. And as soon as the particles cease to be compressed by the striking, then {because flints are rigid, their parts strive to assume their original figure; and} because flints are friable, {this elastic force brings it about that} some of these particles fly apart and, by intermingling with the pure matter of the first element which surrounds them, form a fire. So, if A is a flint[59] with the globules of the second element illustrated among the particles of its surface, B will represent the same flint while it is struck by some hard body and its pores have been narrowed and can contain nothing other than the matter of the first element; while C will represent the same flint after it has already been struck, when certain of its particles have separated from it and, surrounded solely by the matter of the first element, have been transformed into sparks of fire.

---

[58] The French of this article substitutes 'brittle' for 'friable', which is certainly more accurate.
[59] See Plate XX.

85.   How fire is struck from dry wood.

If extremely dry wood is struck in the same manner, it does not thus emit sparks; because, since it is not as hard, the first part of it which encounters the striking body is pushed back toward the second and reaches it before this second part begins to be pushed back toward the third. Thus the globules of the second element do not simultaneously leave many of the wood's interstices, but instead leave successively, now from one, and now from another. Yet if this wood is rubbed vigorously enough for a long time, the unequal agitation and vibration of its particles which results from this friction can drive the globules of the second element out of many of its interstices, and at the same time separate these particles from one another, thus transforming them into fire.

86.   How fire is generated by collecting together the Sun's rays.

Fire is also generated with the aid of a concave mirror or a convex lens, which direct many of the Sun's rays toward one place. For although the action of these rays has the globules of the second element as its foundation, it is however a great deal swifter than their usual motion. And since this action is derived from the matter of the first element, of which the Sun is composed; it has sufficient speed to generate fire; and so many rays can be collected together that they can also have enough force to agitate the particles of terrestrial bodies at this same speed, {in which speed the form of fire consists}.

87.   How fire is generated solely by extremely violent motion.

For in fact it does not matter by what cause the terrestrial particles first begin to be very rapidly moved. For even if they previously have been without motion; if they are mingling only with the matter of the first element, they immediately acquire very rapid motion from that fact alone: for the same reason as a ship, tied by no ropes, cannot be in rushing water without immediately being carried along by it. And even though these terrestrial particles may not yet be thus mingling with the first element; if they are simply agitated sufficiently rapidly by any other cause whatever, that alone will cause them to agitate both one another and also the globules of the second element situated around them. And thus these terrestrial particles will immediately {expel the heavenly globules from around

themselves and} begin to mingle with the first element; and will subsequently be maintained in their motion by it. For this reason, all extremely
violent motion suffices to generate fire. And such a motion is usually found
in lightning strikes and whirlwinds, that is to say, when a high cloud rushes
into another lower one and drives out the intermediate air; as I explained in
the [Discourses on] Meteorology.[60]

88.   How fire is generated by the mingling of diverse bodies.

Yet of course this motion alone is scarcely ever the cause of fire in the
atmosphere; for exhalations are almost always mingled with the air, and
their nature is such that they are easily changed either into a flame or at
least into a bright body. And this causes ignis fatuus[61] to be generated
around the Earth, and lightning in the clouds, and shooting and falling
stars high in the air. For it has already been stated that exhalations are
composed of particles which are very slender and divided into many tiny
branch-like particles, which envelop other slightly larger particles derived
from bitter juices or volatile salts. It must also be noted that these small
branches are usually so tiny and so entwined together that nothing other
than the matter of the first element can pass through their interstices.
However, there are other larger interstices between the particles covered by
these small branches, which are usually filled by the globules of the second
element; in which case, the exhalation does not catch fire. But it also
sometimes happens that these interstices are occupied by the particles of
another exhalation or spirit, and that these particles, expelling the second
element from there, only surrender that place to the first element, and being
immediately carried along violently by its motion, burst into flame.[62]

89.   How fire is generated in lightning and in shooting stars.

And indeed in lightning flashes and strikes, the cause which drives a
number of exhalations together is obvious: it is the falling of one cloud
upon another. However, in tranquil air, if one exhalation is at rest and has
been made dense by cold; another exhalation which comes from a warmer
place, or is composed of particles more suited to motion, or else is driven by

[60] See Discourse VII.
[61] Will-o'-the-wisps, or phosphorescent marsh gas.
[62] The account given in the French text seems quite different, stating that the globules of the
second element are driven from the particles of the exhalation by an external pressure.

some light wind, easily enters violently into the pores of the first and expels the second element from them. And if the particles of the first exhalation are not yet joined so closely to one another that they cannot be separated by this violence of the others, that alone is sufficient to cause them to burst into flames {which rapidly consume the exhalation}: I think that shooting stars are ignited in this way.

90.   How fire is generated in those things which shine but do not burn: as, for example, in falling stars.

However, when the particles of an exhalation unite in a body so thick and viscid that they are not separated in this way, they merely emit some light, similar to that which usually appears in rotten wood, in fish preserved in salt, in drops of sea-water, and in other similar things. For if the globules of the second element are driven by the matter of the first, that suffices to create light; as is sufficiently clear from what has been said above. And when the interstices of several terrestrial particles which are joined together are so small as to have room only for the first element; even though the first element may not have enough force to separate them {and thus to burn the body}, it however easily has sufficient force to drive the surrounding globules of the second element by that action which we have said must be taken to be light. And I believe that falling stars are of this kind; for their matter which has fallen to the ground is often discovered to be viscid and tenacious: although of course it is not certain that it was this same viscid matter which had light: for there could have been some slender flame adhering to this matter, {or a more subtle burning matter surrounding it which is usually consumed before reaching the earth}.

91.   How it is generated in drops of sea-water, in rotten wood, and in similar things.

Moreover, it is easy to see how light is generated in drops of sea-water, the nature of which I explained above. Specifically, while the flexible particles of the drops remain intertwined, the rigid and smooth particles are driven out of the drops by the force of a storm or of any other movement; and, agitating rapidly to and fro like darts, easily drive the globules of the second element out of their vicinity, and thus produce light.[63] However, in

[63] This explanation of the phosphorescence of sea-water is almost identical to the one which Descartes gives in the *Discourses on Meteorology*, Discourse III. The French text differs, stating that only the points of the rigid particles protrude from the drops.

rotten wood, in fish which are beginning to be dried, and in such things, I think that light originates solely from the fact that there are in these (while they are thus shining) many pores so narrow that they admit only the first element.

92.   How fire is generated in those things which grow hot but do not give off light: as for example in stored hay.

However, the fact that particles of some spirit or liquid can sometimes generate fire by entering the pores of a hard or even a fluid body, is shown by moist hay which has been shut away anywhere, by {quick} lime sprinkled with water, by all fermentations, and by not a few liquids known to Chemists, which become hot and sometimes even ignite when mixed. Indeed, there is no other reason why new-mown hay, if shut away before it is dry, should gradually become hot and spontaneously generate a blaze except that many spirits or juices which are accustomed to flowing through the pores of green grasses from their roots to their tips {to serve as nourishment} (and which have paths adapted to their size) remain for some time in the mown grasses. Thus, if these grasses are meanwhile confined in a small space, the particles of those juices, migrating from some blades of grass into others, find many pores in those blades which are beginning to dry which are slightly too narrow to admit both these juices and the globules of the second element. Therefore, while they are flowing through such pores, the juices are surrounded solely by the matter of the first element; and, having been very rapidly driven by the latter, they acquire the agitation of fire. Thus, for example, if the space between the two bodies B and C[64] represents one of the pores of some blade of green grass, and if the slender cords 1, 2, 3, surrounded by tiny globes, are taken to be particles of juices or spirits which are accustomed to be transported by the globules of the second element through pores of this kind; and if, on the other hand, the space between bodies D and E is another narrower pore of a blade of grass which is beginning to dry; the same particles 1, 2, 3, upon entering this narrower pore, can no longer have the second element around them, but only the first. It is clear, therefore, that these particles must follow the moderate movement of the second element when between B and C; whereas between D and E, they must follow the very rapid movement of the first. Nor does it matter that only a very tiny quantity of this first element is found around the particles. Indeed, it suffices for them all to float upon the

[64] See Plate XXI, Fig. i.

first element {in such a way that they are not restrained by the second element or by another body}. In the same way, we see that a ship floating down a river follows its current as easily where it is so narrow that the ship almost touches its banks on both sides, as where it is very wide. However, having been thus rapidly moved {by the first element}, particles 1, 2, 3, have much more force to agitate the particles of surrounding bodies than does the first element itself: just as a ship, striking against a bridge or other obstacle, shakes it more strongly than does the water of the river by which the ship is being carried along. And for that reason, by striking against the harder particles of hay, particles 1, 2, 3, easily separate these from one another, especially when several such particles simultaneously strike the same particle [of hay] from different directions. And if such particles separate and carry along with them a sufficiently great number of particles of hay in this way, a fire occurs. However, when the particles of liquid merely agitate the particles of hay, and do not yet have the force to simultaneously separate many of them from one another, they merely heat and rot the hay slowly, {so there is then in it a sort of fire without light}.

93.   How fire is generated in {quick} lime which has been sprinkled
                 with water, and in the remaining things.

   Similarly, we can believe that when a stone is roasted thoroughly to form lime, {the fire's action drives out some of its parts of the third element, so that} many of its pores which were formerly penetrable only by the globules of the second element are widened to such an extent that they admit particles of water; however, these are surrounded only by the first element. In order that I may explain everything here simultaneously; whenever some hard body grows hot by the admixture of some liquid, I judge that this occurs because many of the hard body's pores are of such a measurement as to admit the particles of this liquid, surrounded solely by the matter of the first element. Nor do I think that there is a different reason when one liquid is poured into another liquid: for one or the other is always composed of branching particles entwined and connected in some way, and thus takes the place of the hard body [just mentioned]: as was stated a little earlier[65] concerning exhalations specifically.

[65] See Article 89.

94.   How fire is ignited in the earth's cavities.

However, fire can be ignited in all these ways, not only upon the surface of the earth, but also in its cavities. For there, acrid spirits can penetrate the pores of dense exhalations in such a way as to kindle a flame in them; and fragments of rocks or flints, eroded by a concealed fall of water or by other causes, can fall down from the vaulted roofs of the cavities onto the floor beneath and both expel the trapped air with great force and also generate fire by the striking of flints. And when a body has once started a blaze, it is easily also transmitted to other neighboring flammable bodies. For the particles of flame, upon encountering the particles of these bodies, move and carry off with them these other particles. However, this is less relevant to the generating of fire than to its maintenance, which must next be discussed.

95.   How a candle burns.

For example, let us consider the lighted candle AB,[66] and let us think that in the whole space CDE which is occupied by its flame, many particles of wax, or of whatever other oily matter this candle is made, are flying to and fro; and also many globules of the second element. But [let us think] that both the globules and the particles of wax are floating upon the matter of the first element in such a way as to be carried rapidly along by its motion; and that although these often touch and drive one another, they do not however support one another entirely, as they are accustomed to do in places where there is no fire.

96.   How fire is maintained in it.

However, the matter of the first element (which is found in great abundance in this flame) is always striving to leave the place in which it is, because it is very rapidly moved; and indeed to leave in an upward direction, that is, to recede from the center of the Earth; because, as was stated above, it is lighter than those heavenly globules occupying the pores of the air. And both those globules, and also all the terrestrial particles of the surrounding air, are striving to descend into its place; and thus they would immediately stifle the flame if it were composed solely of the first

[66] Articles 95 through 98 refer to Plate XXI, Fig. ii.

element. But the terrestrial particles which constantly leave the wick FG are instantly immersed in the first element and follow its course. And when they encounter those particles of air which were ready to descend into the place of the flame; they repulse them and thus maintain the fire.

97.   Why its flame is pointed and emits smoke.

Since these particles {follow the course of the first element, they} mainly strive to ascend, {and} the flame is as a result usually pointed. And because these particles are much more rapidly moved than those particles of air which they repulse, they cannot be prevented by the latter from proceeding toward H, where they gradually lose their agitation and are thereby transformed into smoke.

98.   How air and other bodies feed a flame.

Because there can be no void, this smoke would find no place in all the air if, as it left the flame, an equal quantity of air did not return in a circular movement toward the flame. Specifically, when the smoke ascends to H, it drives the air out of there toward I and K; and this air, skimming the tip B of the candle and the roots F of the wick, reaches the flame and serves to feed it. But air {alone} would not suffice for this effect (because of the smallness of its parts) if it did not bring with it, via the wick, many particles of wax agitated by the heat of the fire. And thus the flame must be constantly renewed in order to be maintained; and does not remain the same, any more than does a river into which new waters are always flowing.

99.   Concerning the movement of air toward a fire.

Now this circular movement of air and smoke can be observed whenever a large fire is kindled in some bedchamber. For if the bedchamber is closed up in such a way that (apart from the stove-pipe through which the smoke is leaving) only one aperture of some kind exists; a great draft, coming through this aperture toward the stove to take the place of the departing smoke, will be continuously felt.

100.   Concerning those things which extinguish fire.

From the preceding, it is obvious that two things are required for the conservation of fire. First, there must be terrestrial particles in it which have the force, when they have been driven by the first element, to prevent the fire from being stifled by air or by other fluids situated above it. I am speaking only of fluids situated above the fire ; because, since only their weight carries them toward the fire, there is no danger that the fire could be extinguished by those fluids which are below it, {and which approach the fire only when drawn toward it to feed it}. Thus, the flame of an inverted candle is extinguished by a liquid which at other times maintains it. On the other hand, fires can be created in which the terrestrial particles are so solid, so numerous, and above all, agitated with such great impetus, that these fires repulse even water which has been poured on them ; so that they cannot be extinguished by water.

101.   What is required in order for a body to be suited to feeding a
              fire.

The other thing which is required for the maintenance of fire is that it must adhere to some body from which new matter can {continuously} come to it, to replace the departing smoke. Therefore that body must have in it many particles which are sufficiently tiny in proportion to the fire to be maintained. And these particles must be joined to one another or to other bulkier ones in such a way that they can be separated, by the violence of the particles of this fire, from one another and from the neighboring globules of the second element, and thus be transformed into fire.

102.   Why a flame from spirits of wine[67] does not burn a linen cloth.

I say that the particles of this body must be sufficiently tiny in proportion to the fire to be maintained ; because, for example, if spirits of wine which have been sprinkled onto a linen cloth catch fire, this very slender flame will indeed consume all these spirits of wine ; but it will not affect the linen cloth, which another fire would easily burn up : since the particles of cloth are not sufficiently tiny for this fire to be able to move them.

---

[67] That is, alcohol distilled from wine.

103.   Why spirits of wine ignite very easily.[68]

And indeed spirits of wine very easily feed a flame, because they consist only of extremely slender particles and because these [particles] contain certain tiny branches which in fact are so short and flexible that they do not adhere to one another (for in that case the spirits would be transformed into oil): but which leave around themselves many very small spaces which cannot be occupied by the globules of the second element, but only by the matter of the first.

104.   Why water is very difficult [to ignite].

Water, on the other hand, seems to be very much opposed to fire, because it is composed of particles which are not only fairly bulky, but also perfectly smooth: consequently, nothing prevents the globules of the second element from surrounding these particles on all sides, {leaving very little space for the first element}, and following them. In addition, water is composed of flexible particles; as a result, it easily enters the pores of bodies which are being burned, and, by driving particles of fire out of these pores, prevents other particles of the bodies from starting to burn.

105.   Why the force of great fires is increased by water or by salts
            thrown upon them.

Yet {that depends on the ratio between the size of its parts and the violence of the fire or the size of the pores of the burning body}. Some bodies are such that particles of water introduced into their pores help the fire; because the particles spring violently out of these pores and are themselves ignited.[69] Therefore smiths sprinkle fossil[70] coals with water. Moreover, a small quantity of water, thrown upon huge flames, increases them. Salts accomplish this effect even more powerfully: for their particles, being rigid and oblong, are agitated in the flame like little darts, and have great force to agitate the tiny particles of other bodies against which they

---

[68] The French text of this article is much longer. The additional material refers to Article 102 and contains a long description and explanation of the circumstances in which the cloth will or will not be ignited by an alcohol flame; stating that it will fail to ignite only if it remains moistened with the alcohol.

[69] The French text omits the claim that the particles of water are ignited.

[70] The Latin term is *'fossilis'*, which simply means "dug up" or "mined".

strike: as a result, it is customary to add salts to metals in order to melt them {more easily}.

106.   The nature of bodies which are easily burned.

However, those bodies which are generally used as fuel for fire, such as woods and other similar things, are composed of various particles. Certain of these are very tiny, while others are slightly bulkier, and the rest are of gradually increasing size. Most of these particles are branching, with great pores lying between them. As a result, the particles of fire which have entered these pores very rapidly move first the tiniest particles of these bodies, next the medium-sized ones, and with their help, the bulkiest. And thus the fire first drives the heavenly globules out of the narrowest interstices, and then out of the rest; and carries off with it all these particles except the bulkiest, from which ashes are formed.

107.   Why certain [burning] bodies burst into flame, and others do not.

And when particles of this kind, which simultaneously leave a burning body, are so numerous that they have the force to expel the heavenly globules from some space in the nearby air, they fill that space with flame. If, however, they are less numerous, there occurs a fire without flame, which either[71] gradually creeps through the pores of its tinder, when the matter which it can feed upon is found there: as in those [slow] fuses or wicks which are used in time of war to ignite gunpowder in engines for hurling missiles.

108.   Why fire endures for a time in live coals.

Or, if the fire has no such material around it, it is not maintained; except in so far as it is imprisoned in the pores {between the larger and unignited particles} of the body to which it adheres, and needs a certain amount of time to separate all the particles of that body in order to free itself from them. And this can be seen in ignited coals covered by ashes; which retain fire for many hours, solely because that fire is in certain tiny branching particles which are entwined with other bulkier ones and cannot escape except a few at a time, even though they are very rapidly agitated. And,

[71] The second part of this disjunction is the first sentence of Article 108.

perhaps, before they may thus leave, these tiny particles must be worn away by prolonged motion, and individual ones divided into several others.

109.   Concerning gunpowder[72] manufactured from sulphur, niter, and charcoal; and first, concerning sulphur.

Yet nothing catches fire more rapidly, or retains it for a shorter time, than does gunpowder made of sulphur, niter, and charcoal. For in fact, sulphur is by itself as flammable as possible; because it is composed of particles of acrid juices, which are coated by such tiny and dense branches of oily matter that very many pores, between these branches, are accessible only to the first element. As a result, sulphur is also considered extremely hot for medicinal purposes.[73]

110.   Concerning niter.

Niter, on the other hand, is composed of oblong and rigid particles, which however differ from those of common salt, in that they are thicker at one end than at the other. This is obvious, for example, from the fact that when niter has been dissolved in water, it does not congeal on its surface in a square figure, as common salt does, but adheres to the bottom and sides of the flask,[74] {and thus its parts must be larger or heavier at one end than at the other}.

111.   Concerning the combination of sulphur and niter.

And, as for the size of the particles {of niter and sulphur}, it must be thought that the proportion between them is such that those of the acrid juices in sulphur (when they have been set in motion by the first element), very easily drive the globules of the second element out of the interstices between the branches of oily matter; and at the same time they violently agitate the particles of niter, which are larger than those of sulphur.

---

[72] Gunpowder is composed of potassium nitrate, charcoal, and sulphur; in varying proportions. In Descartes's time, the formula would have been roughly sixty-six percent potassium nitrate, seventeen percent charcoal, and seventeen percent sulphur.

[73] Galen, among many others, makes such a claim.

[74] In the third discourse of the *Meteorology*, Descartes claims that common salt remains dissolved in sea-water because its particles are not larger at one end than at the other and that if they were, they would fall to the bottom of the sea.

112.   Concerning the motion of the particles of niter.

And these particles of niter point downward at that end at which they are
thicker and therefore also heavier. And for that reason, their principal
movement is in the sharper end which, pointing upward as at B,[75] is
agitated in a circle which is tiny at first, as at C; but which (unless prevented
by something) immediately becomes larger, as at D. Meanwhile, the
particles of sulphur, {which do not whirl around in the same way}, having
spread very rapidly {in straight lines} in all directions, reach the other
particles of niter in an extremely short time, {and ignite them by expelling
the second element from around them}.

113.   Why the flame of this powder expands greatly, and principally
              acts on things above it.

And the fact that these particles of niter each require much space in order
to describe the circles of their movement, causes the flame of this powder to
expand exceedingly. And because they describe these circles with that
pointed end which is held upward, all their force consequently tends toward
things situated above them; and when gunpowder is very dry and fine, it
can harmlessly be ignited in the hand.

114.   Concerning charcoal.

Moreover, charcoal is mixed with sulphur and niter, and grains or
globules are formed from this mixture after it has been sprinkled with some
liquid; and these are then thoroughly dried {to form gunpowder}. There are
in fact many pores in charcoal. First, because pores were previously very
numerous in the bodies from which the charcoal has been formed by
burning {and then extinguishing the fire before the wood has completely
burned}. Second, because while these bodies were being burned, {many
particles left them and thus} a great deal of smoke ascended from them.
And only two kinds of particles are found in charcoal: one kind are fairly
large ones which {could not be converted into smoke and which}, when
alone, form ashes {if the charcoal is permitted to burn completely}. The
others are smaller and of course ignite very easily, because they have
already been agitated by the force of fire. But, being entwined by long and
complex branches, they cannot be detached without some force: as is

[75] See Plate XXI, Fig. iii.

evident from the fact that, while the other particles were leaving the preceding fire in the form of smoke, these remained last.

115.   Concerning the grains of this powder, and in what the force of this powder principally consists.

Thus {the particles of} sulphur and niter easily enter into the wide pores of charcoal, and are coated and bound together by its branching particles; especially when this mixture has been moistened with some liquid, formed into grains or granules, and afterwards dried. And the purpose of this procedure is to cause many of the particles of niter to ignite, not merely one at a time, but simultaneously.[76] For when fire (which has been brought from elsewhere) first touches the surface of some grain, it does not immediately ignite and destroy that grain; rather, it requires a certain time to penetrate from the surface into the interior parts of the grain. And after first igniting the sulphur there, the fire gradually also causes violent agitation in the particles of niter; so that, when these particles have finally acquired force, and demand a great space in which to describe their circles, they break the bonds of the charcoal and shatter {and ignite} the entire grain. And although this time is extremely short if compared to hours or days; it must be noted that it is quite long if compared with that great speed at which the exploding grain spreads its flame throughout all the neighboring air. When, for example, in a military engine for hurling missiles, certain grains of powder are first ignited by the fire of the fuse or other primer; the flame erupting from them spreads instantly through all the interstices of the surrounding grains. And although this flame cannot so suddenly penetrate into their interior parts, yet because it touches many grains at the same time; it causes these to ignite and expand simultaneously, and thus fire the weapon with great force. Thus the resistance of the charcoal greatly increases the speed of the particles of niter when they burst into flame. And granulation is necessary so {that the size of the grains and the amount of charcoal is proportionate to the size of the gun and so} that there may be, around these granules, spaces sufficiently large to permit the flame of the initially ignited powder free access to the many parts of the remaining powder.

---

[76] In fact, it is the sulphur and charcoal which burn; the potassium nitrate furnishes large quantities of oxygen, greatly increasing the combustion rate.

116.   Concerning oil-lamps which have remained burning for a very
            long time.

After that fire [of gunpowder], which is the least enduring of all, let us
consider whether any other can exist which, on the contrary, would last for
a very long time without any fuel: as is narrated concerning certain oil-
lamps, which have sometimes been found burning after [being for] many
years in underground vaults where the bodies of the dead are preserved. {I
do not wish to vouch for the truth of such stories, but it seems to me that} in
a subterranean and very tightly closed place (where the air is either never
moved by any winds, or moved only by the slightest ones); it could perhaps
have occurred that many branching particles of soot accumulated around
the flame of the lamp, and that these particles, resting upon one another,
remained motionless and thus formed a sort of small dome, sufficient to
prevent the surrounding air from overwhelming and stifling the flame. And
these particles may also have sufficed to break and dull the force of this
same flame, so that it could not ignite any more particles of the oil or the
wick, if any still remained. As a result, the matter of the first element
remained there alone, and constantly gyrated very rapidly, as if in some
tiny star; and it thereby repulsed from around itself the globules of the
second element (to which alone passage was still open between the particles
of soot accumulated around the flame),[77] and thus it spread light through
the entire tomb. A tiny and dim light indeed, but one which could easily
recover its force from the movement of the external air when the place was
opened up and the soot had been dispersed; and a burning lamp would thus
be revealed. {Although the lamp may perhaps go out soon afterward,
because this flame was probably only able to maintain itself in this way
after having consumed all its oil}.

117.   Concerning the remaining effects of fire.

Now, let us consider those effects of fire which have not yet been able to
be learned from the manner in which fire occurs and is maintained. For in
fact, from what has been said, it is now obvious how fire gives off light, how
it heats, and how it breaks down all the bodies on which it feeds into many
particles. And it is also now clear how the particles leave such bodies; first
the most slender and slippery ones, next others which may perhaps be no

[77] The French text has, "(which alone still strive to approach the flame through the pores
which they have preserved for themselves in this dome)" here.

bulkier than the first but are more branching and entwined (namely, those which adhere to the sides of chimneys and form soot); and how only the bulkiest of all remain as ashes. However, it remains for us to show briefly how, by the force of fire, certain bodies by which it is not fed become liquid and boil, while others are dried and become hard; [and how] some are given off as exhalations, and some transformed into lime or glass.

118.   Which bodies become liquid and boil after being placed near
            fire.

All hard bodies become liquid while enduring the force of fire if they are composed of particles which are separated from their neighbors with more or less equal difficulty, and which can be disunited by some force of fire. For to be fluid is nothing other than to be composed of particles which are separated from one another and are in some [individual] motion. And when the agitation of these particles is so great that some of them are transformed into air or fire, and require more space than usual for their motion, and thereby drive up [some of] the others; this fluid body bubbles and boils.

119.   Which bodies are dried and become hard.

However, when bodies which contain many slender, flexible, and slippery particles (which are entwined with but not very firmly joined to other bulkier or branching ones) are brought near to a fire; they give off these slender particles, and by that sole fact become dry. For to be dry is simply to lack those fluid particles which form water or another liquid when gathered together. And while these fluid particles are enclosed in the pores of hard bodies, they dilate those pores, and, by their motion, agitate the other particles of these bodies. This removes, or at least diminishes, the bodies' hardness. But when these fluid particles have been given off, the others which remain are usually more closely joined and more firmly connected, and thus the bodies become hard.

120.   Concerning ardent, insipid, and acid waters.

Moreover, the particles which are thus given off are differentiated into various types. First, let me disregard those which are so mobile and slender that they can form no body except air when not mixed with other bodies. The slenderest of all the others are also very easily given off; and when captured in Chemists' flasks perfectly sealed on all sides, and gathered together, they form ardent waters, or spirits; such as are usually extracted from wine, wheat, and many other bodies. Then follow the sweet or insipid waters, such as those which are distilled from plants or other bodies. Third, there are the eroding and acid waters, or bitter juices, which are extracted from salts; though not without great force of fire.

121.   Concerning sublimates and oils.

In addition, certain bulkier particles (such as those of quicksilver and of salts, which adhere to the tops of [heated] flasks and congeal into solid bodies) need fairly great force in order to be sublimated. However, oils are given off by hard and dry bodies with the greatest difficulty of all; and this [task] must be performed, not so much by force of fire, as by a certain skill. For, inasmuch as the particles of oils are slender and branching, great force would break and destroy them {entirely changing their nature}, before they could be drawn out of the pores of these bodies. Instead, an abundance of water is poured onto these bodies, and (since water's particles are smooth and slippery), as they flow through those pores, they gradually extract the oily particles, and carry them along intact.

122.   That when the degree of fire has been changed, its effect is changed.

And in all these distillations, the degree of fire must be observed: for when it has been altered, the effect is always altered in some way. Thus, when exposed first to a slow fire and then to a gradually stronger one, many bodies become dry and give off various particles. But they would not give off such particles, but would rather become liquified in their entirety, if they were subjected to intense heat from the beginning.

123.   Concerning lime.

The method of applying fire also varies its effect. Thus, certain bodies become liquid if they grow hot all at once, as a whole. Whereas if a strong flame licks their surface, it converts that surface into lime. For in fact, all hard bodies which are reduced to a very fine powder solely by the action of fire (that is, when certain of their smaller particles, which were joining the rest together, have either been broken or expelled) are commonly said by

Chemists to be converted into lime {or calcified}. Nor is there any other difference between ashes and lime, except that ashes are the remains of those bodies of which a great part has been consumed by fire, while lime is the remains of those bodies which are left almost intact after the combustion is finished, {losing only those few parts which served to unite the others}.

124.   Concerning the way in which glass is made.

The final effect of fire is the transformation of lime and ashes into glass. For after all the smaller particles have been plucked or driven out of bodies which are burning, the particles which remain to form lime or ashes are so solid and bulky that they cannot be carried upward by the force of fire; and they have, for the most part, irregular and angular figures. As a result, though they rest upon one another, they do not adhere to one another, and are not even contiguous to one another except perhaps at certain very tiny points. However, when a strong and enduring fire subsequently exerts its force vigorously against them; (that is, when the smaller particles of the third element and the globules of the second, which have both been carried along by the matter of the first element, are vigorously and very rapidly moved around these particles of lime and ashes on all sides); their angles gradually become blunted, and their surfaces become smooth, and perhaps some of them are also bent. And thus, by crawling and flowing over one another, they come to touch one another, not only at points, but on certain tiny surfaces; and having been joined together in this way, they form glass.

125.   How the particles of glass are joined together.

For in fact, it must be noted that when two bodies whose surfaces have some width approach each other along a straight line, they cannot approach so closely that there is not some space between them which is occupied by the globules of the second element. However, when one body is led, or crawls, obliquely over the other, they can be much more closely joined. For example, if bodies B and C[78] approach each other along straight line AD; the heavenly globules, trapped by the surfaces of these bodies, will prevent their immediate contact. However, if body G is moved back and forth above body H along straight line EF, nothing will prevent G from

[78] See Plate XXI, Fig. iv.

being immediately contiguous to H; at least if the surfaces of both are smooth and plane: however, if they are rough and uneven, they will gradually become smooth and plane by means of that very motion. Therefore, it must be thought that the particles of lime and ashes, separated from one another, are here illustrated by bodies B and C; while the joined particles of glass are illustrated by bodies G and H. And, from this one difference {in their manner of joining}, which it is obvious must be induced in them by the violent and prolonged action of fire, {we can understand perfectly and explain how} these particles acquire all the properties of glass.

126.   Why it is fluid when glowing with heat, and why it easily assumes all figures.

For glass, when still glowing with heat, is fluid, because its particles are easily moved {separately from one another} by that force of fire which previously smoothed and bent them. However, when it begins to be cooled, it can take on any figures whatever. And this is common to all bodies which have been liquified by fire; for while they are still fluid, their particles effortlessly adapt themselves to any figures whatever, and when such bodies subsequently harden with cold, they retain the figures which they last assumed, {since cold halts the movement of their parts}. Glass can also be drawn out into slender threads like hairs, because when its particles are just beginning to harden, they flow over one another more easily than they separate from one another.

127.   Why glass is extremely hard when it is cold.

Then, when glass has completely cooled, it is extremely hard but at the same time also extremely fragile; and it is the more fragile, the more rapidly it has been cooled. And the cause of this hardness must be that glass consists entirely of fairly large and inflexible {hard} particles, {which fire cannot break and} which adhere to one another by immediate contact rather than by the interweaving of tiny branches. For most other bodies are softer because their particles are flexible, or at least end in certain small flexible branches which join the particles to one another by their mutual interlacing. However, no joining of two bodies can be stronger than that which arises from their immediate contact; that is to say, when they touch each other in such a way that neither is in motion to separate itself from the other.[79] This happens to the particles of glass as soon as they have been removed from the fire; because their bulkiness, their contiguity, and the inequality of their figures make it impossible for the surrounding air to maintain in them that motion by which they were separated from one another {by the fire}.

128.   Why glass is extremely fragile.

Yet glass is nevertheless extremely fragile, because the surfaces along which its particles touch one another are few and very narrow. And many other softer bodies are more difficult to break, because their parts are interwoven in such a way that they cannot be separated unless many of their tiny branches are broken and torn away. {And there are many more particles which must be disunited in order to break these softer bodies than there are little surfaces to separate in [order to break] glass.}

129.   Why its fragility is decreased if it is cooled slowly.

It is also more fragile when it is cooled quickly than when it is cooled slowly; for its pores are fairly open while it is glowing with heat, because then much matter of the first element, together with globules of the second. and also perhaps with some of the smaller particles of the third, passes through these pores. However, when glass cools naturally, these pores become narrower; because [then] only the globules of the second element are passing through them, and this requires less space.[80] And if the cooling occurs too rapidly, the glass becomes hard before its pores can thus contract: as a result, those globules subsequently always make an effort to separate its particles from one another.[81] And since these particles are joined together solely by their own contact, one cannot be even very slightly separated from another without many neighboring ones on the surface on which this separation began to occur also being immediately separated, and the glass being thereby completely broken. For this reason, those who make glass vessels remove them gradually from the kilns, so that they may

[79] See Part II, Article 55.
[80] This explanation of the narrowing of the pores is omitted in the French text.
[81] The French text reads, "If it cools too rapidly, its parts do not have time to dispose themselves in such a way that the pores can all contract equally, so the matter of the second element, which subsequently passes through these pores, strives to make them equal and thus breaks the glass."

cool slowly. And if a cold glass vessel is placed near a fire in such a way that it grows much hotter in one area than in nearby ones, that alone will break it in that area: because the pores of the heated area cannot be dilated, and the pores of the nearby areas remain unchanged, without that area being separated from the others. But if the glass vessel is first placed close to a slow fire, and then to a gradually more intense one, and if it grows hot in all areas equally; it will not break: because all its pores will be dilated equally and simultaneously.

130   Why it is transparent.

In addition, glass is transparent because it is a liquid while it is being formed, and the matter of fire flows around its particles on all sides and hollows out for itself innumerable pores. The globules of the second element subsequently pass freely through these, and can transmit the action of light in all directions along straight lines. For it is not necessary for this transmission that these pores be perfectly straight, but only that they be nowhere interrupted. So, for example, if we imagine glass to consist of particles which are perfectly spherical and equal, but so bulky that the globules of the second element can pass through that triangular space which must remain between three globules which are contiguous to one another; that glass would be completely transparent;[82] although much more solid than any glass which now exists.

131.   How it is colored.

However, when the materials from which glass is made are mixed with metals or other bodies whose particles are more resistant to fire and less easily made smooth than are those other particles which compose glass: that suffices to make the glass less transparent and assume various colors; depending on the degree to which, and the diverse ways in which, these harder particles block its pores. {And this [partial blockage] causes the parts of the second element which are passing through these pores to rotate in various ways; and it is this rotation which causes colors as I have shown in the [Discourses on] Meteorology}.[83]

[82] See Article 17.
[83] See Discourse VIII.

132.   Why it is rigid[84] like a bow; and, in general, why rigid things,
               when they have been bent, spontaneously return to their former
               figure.

Finally, glass is rigid: that is to say, it can be somewhat bent by external
force without breaking but afterwards springs back violently and re-
assumes its former figure, like a bow. This is clearly seen when it has been
drawn out into very slender threads. And the property of springing back in
this way generally exists in all hard bodies whose particles are joined
together by immediate contact rather than by the entwining of tiny
branches. For, since they have innumerable pores through which some
matter is constantly being moved (because there is no void anywhere), and
since the shapes of these pores are suited to offering free passage to this
matter (because they were earlier formed with its help), such bodies cannot
be bent without the shapes of these pores being somewhat altered. As a
result, the particles of matter accustomed to passing through these pores
find there paths less convenient than usual and push vigorously against the
walls of these pores in order to restore them to their former figure. For of
course if, for example, those pores through which the globules of the second
element are accustomed to pass are circular in an unbent bow, it must be
thought that these pores are elliptical in a bow which is taut or bent; and
that the globules which strive to pass through these pores strike against
their walls at the smaller diameters of these ellipses, and thus have the force
to restore them to a circular shape. And although this force is very tiny in
the individual globules of the second element, the united and concerted
force of all the very many globules which constantly strive to pass through
the numerous pores is sufficiently great to restore the bow to its former
shape. However, a bow which has remained bent for a long time (especially
if it is made of wood or of another material which is not extremely hard)
gradually loses the force to spring back: this is due to the long period of
wearing away of [the walls of] these pores by the particles of matter passing
through them, which gradually better adapts their shape to the size of those
particles.

133.   Concerning the magnet.[85] A repetition of those of the things
               previously said which are necessary for an explanation of it.

So far, I have attempted to explain the natures, and also the principal

---

[84] Both the Latin and French have 'rigid' here, although 'elastic' is clearly what is intended.
[85] By 'magnet', Descartes means a loadstone; that is, a naturally magnetized piece of iron ore.

powers and qualities, of air, water, earth, and fire, which are commonly thought to be the elements of this globe which we inhabit, {because these are the bodies most generally found there}; it now remains for me to speak of the magnet. For inasmuch as its force is spread throughout this whole globe of the Earth, {whose entire bulk is in fact a magnet}; there is no doubt that this subject is relevant to a general consideration of the Earth. Accordingly, we shall now recall to mind those grooved particles of the first element, which were quite carefully described above in Article 87 of Part III, and in the articles following. Moreover, understanding all that was said in Articles 105 to 109 [of Part III] concerning star I to here apply to the Earth, we shall think that there are many pores in the Earth's intermediate region which are parallel to its axis, and through which the grooved particles coming from one pole freely proceed to the other. And these pores have been hollowed out to the measurement of these particles in such a way that those which accept the grooved particles coming from the South pole can in no way admit those which come from the North pole; conversely, those which accept the Northern particles do not admit the Southern ones: because of course these particles are twisted like the thread of a screw; some in one direction and the others in the opposite direction. Furthermore, [we shall remember that] the same particles can enter through only one end of these pores, and cannot return through the opposite one because of certain extremely tiny extremities of branches in the windings of these pores, which have been bent in that direction in which the grooved particles are accustomed to pass, and which spring back in the opposite direction in such a way as to prevent their return. As a result, after these grooved particles have traversed the whole intermediate Earth from one hemisphere to the other along straight lines, or lines equivalent to straight, parallel to its axis; they return through the surrounding aether to that same hemisphere through which they earlier entered the Earth; and thus flowing through the Earth again, form a kind of vortex.

134.   That there are no pores either in air or water suited to admitting
              the grooved particles.

And, since we have shown that four different bodies can have been created out of that aether through which we had said that the grooved particles return from one pole to the other (namely, the interior or metallic crust of the Earth, water, the exterior earth, and air); and since we have noted in Article 113 of Part III, that no vestiges of the pores formed to the

measurement of the grooved particles can have been preserved except in the
bulkier particles of this aether: it must be noticed here that all these bulkier
particles flowed together in the beginning to the interior crust of the Earth.
And there cannot be any such pores in either water or air: both because
there are no sufficiently large particles there; and also because since these
bodies are fluid, their particles are constantly changing situation; and
therefore, any such pores which might have once been in these bodies
would long ago have been destroyed by this changing, since such pores
require a certain and fixed situation.

135.   And that there are none in any bodies of the exterior earth,
         except in iron.

In addition, since it was stated before that the interior crust of the Earth
consists partly of branching particles joined to one another, and partly of
others which are moved this way and that through the interstices of the
branching ones; these pores cannot exist in the latter more mobile particles
either, for the reason just indicated, but only in the branching ones. And as
for the exterior earth, there certainly were no such pores in it either in the
beginning, since it was formed between the water and the air {and thus was
composed of very small particles}. But since various metals subsequently
ascended from the interior earth to this exterior one, and even though
all those metals which are composed of the more mobile and solid particles
of that interior earth must not contain pores of this kind; certainly that
metal which is composed of particles which are branching and bulky but
not so very solid, cannot lack these pores. And it is thoroughly in conformity with reason for us to believe iron {or steel} to be such a metal.

136.   Why there are such pores in iron.

For no other metal is so difficult to shape with a hammer {without the aid
of fire}, or to melt over fire, nor can any be made so hard without the
admixture of another body: which three things serve as proof that its
scrapings are more branching or angular than those of other metals, and
consequently more firmly joined together. Nor is this contradicted by the
fact that some lumps[86] of iron [ore] melt fairly easily over fire the first time

[86] The term used here is 'gleba', which means "a clod of earth"; thus, Descartes is referring
here and elsewhere to pieces of iron ore.

{after they have been mined}; for then their scrapings are not yet joined together but are separate, and therefore are easily agitated by the force of heat. Besides, although iron is harder and less fusible than other metals, it is nevertheless one of the least heavy, and is easily damaged by rust or eroded by *aqua fortis*:[87] all of which things serve as proof that its particles are not more solid in proportion to their greater size[88] than those of other metals, and that they contain many pores.

137.   How there also are pores in its individual scrapings.

However, I do not wish to affirm here that there are whole tunnels[89] (twisted like the thread of a screw, through which pass grooved particles), in the individual scrapings of iron; nor do I wish to deny that many such tunnels may be found in these scrapings: but it will suffice here for us to think that halves of such tunnels are hollowed out in the surfaces of individual scrapings, so that they form whole tunnels when these surfaces are appropriately joined. And it can easily be believed that those bulkier branching and hollowed-out particles of the interior earth, from which iron is formed, were divided by the force of spirits or bitter juices permeating the interior earth in such a way that these halved tunnels remained on the surfaces of the scrapings which were being divided; {because when a hard body with many round holes is broken, it usually divides along lines which pass directly through those holes}. And subsequently, these scrapings ascended little by little into mines through the veins of the exterior earth, thrust out both by spirits and also by exhalations and vapors.

138.   How these pores are made suitable to admitting the grooved particles coming from either direction.

And it must be noted that these scrapings [of iron] cannot always be turned in the same directions as they are thus ascending; because they are angular and because they strike against diverse inequalities in the veins of the earth. Also, when the grooved particles (which emerge violently from the interior earth and seek paths for themselves throughout the whole exterior earth) find the pores of these scrapings so situated that in order to

[87] *Aqua fortis* is nitric acid.
[88] That is, the particles of iron are less dense than those of other metals.
[89] The term used in this article is '*foramen*', which means 'an opening made by boring'; elsewhere, Descartes always uses '*meatus*' or '*porus*', which means "passage" or "pore".

continue their movement along straight lines they must strive to enter through those pore openings from which they were formerly accustomed to leave; they encounter those very tiny extremities of branches which, as stated above, protrude among the windings of the pores and rise up against the grooved particles which are about to return. We must note, too, that these extremities of branches initially do indeed resist these grooved particles; but after having been very frequently struck by them, these extremities are eventually all bent in the opposite direction, or else some of them are broken. And subsequently, when the position of the scrapings in which the pores are situated has changed (and the pores accordingly turn their other openings toward the grooved particles), the latter again encounter the extremities of the branches springing up in the pores, and again gradually bend them in the other direction; and the more frequently and the longer this process is repeated, the easier it becomes to bend these branches in both directions.

139.   What the nature of the magnet is.

And those scrapings which have been frequently turned, (sometimes in one direction, sometimes in another) while thus ascending through the veins of the exterior earth, whether they have accumulated alone or have been driven into the pores of other bodies, form a lump of iron. On the other hand, those which have either always retained the same position, or else, if sometimes forced to change it in order to reach the mines, have at least subsequently remained immobile there for many years after having been firmly driven into the pores of a rock or other body, form a magnet. And thus there is scarcely any lump of iron {ore} which does not in some way approach the nature of a magnet, and there is absolutely no magnet in which there is not some iron contained. Though perhaps this iron may sometimes adhere so closely to some other bodies that it can more easily be damaged than detached from these bodies by fire. {And often the stone in which these magnetic fragments are trapped is very hard, so it is sometimes almost impossible to melt magnets to make iron.}

140.   How steel and any kind of iron are formed by smelting {the ore}.

However, when lumps of iron {ore}, which have been placed near a fire, are liquified in order to be converted into iron or steel; their scrapings, agitated by the force of heat and separated from the heterogeneous bodies, twist this way and that until they can attach themselves to one another along those surfaces in which (as previously stated) the halves of the pores suited to admitting the grooved particles have been hollowed out; and also until the halves of these pores match so perfectly that they form whole pores. When this happens, the grooved particles, which are found in fire no less than in {all} other {terrestrial} bodies, flow through those pores more freely than through other places, and prevent the narrow surfaces (the appropriate situation and union of which re-creates the pores) from changing situation as easily as before. And the contiguity of these surfaces, or at least the force of weight (which pushes all the scrapings downward), prevents these surfaces from being easily separated. And since the scrapings themselves continue meanwhile to be moved by the agitation of the fire; certain groups of them unite in the same [individual] movement, and all the fluid formed by them is divided, so to speak, into various droplets or grains: that is to say, all those scrapings which are moved together form, as it were, one drop; which immediately smoothes and polishes its own surface by its own movement. For, by its encounter with other drops, whatever is rough and angular in the scrapings of which it is composed is thrust from its surface toward its interior parts; and thus all the parts which form each [individual] droplet are joined together extremely closely.

141.　　　Why steel is extremely hard, rigid and brittle.

And all the fluid which has thus been divided into droplets or grains, if it cools rapidly, solidifies into steel, which is extremely hard, rigid, and brittle; almost like glass. Now, it is hard because it consists of scrapings which are very closely joined together; and rigid (that is to say, such that, if bent, it spontaneously returns to its former figure), because the narrow surfaces of its scrapings are not separated by this bending, but only its pores change their figures, as was said earlier concerning glass. Finally, it is brittle because the droplets or grains of which it consists adhere to one another only by the contact of their surfaces; and this contact can be immediate only in extremely few and narrow areas.

142.　　　The difference between steel and other iron.

However, not all iron ores are equally suited to being transformed into steel; and even those from which the best and hardest steel is usually made give only base iron when they are melted by an inappropriate fire. For if the scrapings of which the lumps are composed are so angular and rough that they adhere to one another before they can correctly attach their surfaces to one another and be divided into droplets; or if the fire is not sufficiently intense to thus divide the liquid into droplets and simultaneously compress the scrapings forming them; or, on the other hand, if the fire is so intense that it disturbs the correct position of these scrapings: steel is not formed, but iron, which is less hard and more flexible.

143.   How steel is tempered.

Furthermore, even though steel which has already been made does not easily melt {again and become similar to common iron} if again exposed to fire (because its granules are too bulky and solid to be entirely moved by the fire, and because the scrapings composing each droplet are too firmly joined together to be capable of being entirely expelled from their places): it does nevertheless soften; because all of its particles are shaken by heat. And if it is subsequently cooled slowly, it does not re-assume its former hardness, rigidity, or fragility; but becomes flexible like baser iron. For while it is being cooled in this way, the angular and uneven scrapings which had been driven from the surfaces of the droplets toward their interior parts by the force of heat, {have time to} thrust themselves outward; and, having become entwined with one another like certain very small barbs, bind the droplets to one another. As a result, these scrapings are no longer so firmly joined together in these droplets, and the droplets no longer adhere to one another by immediate contact, but are fastened together as if by certain hooks or barbs. And therefore the steel becomes soft and flexible rather than extremely hard, rigid, and fragile. In this state it does not differ from common iron, except for the fact that such steel recovers its former hardness and rigidity if it has been again heated {red-hot} and then rapidly cooled; while this is not true of iron, or at least not to such an extent. The reason for this is that the scrapings in steel do not move so far away from the situation appropriate for maximum hardness that they cannot easily resume this situation by the force of fire, and retain it in very rapid cooling: however, since the scrapings of iron never had any such situation, they can never resume it {or thus acquire it}.[90] Moreover, in order for hot steel or

[90] The claim being made here, which is made more explicitly in the French text, is that, unlike steel, iron cannot be hardened by heating to red-heat (the temperature at which it becomes malleable) and then quenching. Article 142 clearly implies that some kinds of iron could be made into steel if appropriately remelted.

iron to be thus very rapidly cooled, it is usual to quench it in water or other cold liquids. On the other hand, if it is to be cooled [somewhat] more slowly, it is quenched in oil or other fatty substances. And because the harder and more rigid it is, the more brittle it also becomes; in order to make swords, saws, files, or other instruments from it, it must not always be quenched in the coldest liquids, but in liquids which have been appropriately moderated in temperature depending on whether the extent to which fragility must be avoided in these instruments is greater or smaller than the extent to which hardness must be sought. And therefore, when it is thus quenched in certain liquids, it is not undeservedly said to be tempered.[91]

144.　　　　The difference between the pores of the magnet, of steel, and of iron.

Moreover, as for those pores suited to admitting the grooved particles, it is sufficiently clear from what has been said that they must be very numerous both in steel and in iron; and also that, in steel, they must be more complete and perfect; and that the extremities of the branches protruding in the windings of these pores, when once bent in one direction, cannot so easily be bent back in the opposite direction, although they are indeed more easily bent in steel than in the magnet. And finally, these pores in steel or other iron do not all turn those entrances suited to admitting the grooved particles coming from the South in one direction, and those suited to admitting the other grooved particles coming from the North in the opposite direction, as in the magnet; rather, their situation must be varying and uncertain, because it is disturbed by the action of fire. And in that very brief space of time in which this agitation of fire is halted by cold, only as many of these pores can be turned toward the South and the North as there happen to be grooved particles, coming from the poles of the Earth, which are seeking a passage through them. And because these grooved particles do not correspond in number to all the pores of iron, all iron has received some magnetic force from the situation in relation to the parts of the earth which it occupied when it cooled after its final heating, or else from remaining immobile for a long time in the same situation; but on account of the multitude of pores which it contains, iron can acquire still greater magnetic force [than it acquires in these ways].

[91] Although the rapidity of cooling does make some difference to the hardness of steel, it is usually tempered by being reheated to a much lower temperature after quenching.

145.   An enumeration of the properties of magnetic force.[92]

All of these things follow from the principles of Nature expounded above, in such a way that, even if I were not considering those magnetic properties which I have undertaken to explain here, I nonetheless would not judge these things to be otherwise. However, we shall see, successively, that with the help of these principles, a reason for all those properties {which the most careful experiments . . . have been able to discover up to this time} is furnished so clearly and perfectly that this fact also would seem sufficient to convince us of the truth of these things, even if we did not know that they followed from the principles of Nature. And the magnetic properties which are usually noted by those who admire them can be recorded in the following summaries.

(1) That there are, in a magnet, two poles, one of which everywhere turns toward the North pole of the Earth, while the other turns toward the South.

(2) That, according to the diverse places on the Earth which they occupy, these poles of the magnet incline diversely toward the Earth's center.

(3) That, if two magnets are spherical {and close}, each turns toward the other in the same way as either does toward the Earth.

(4) That after they have thus turned, they approach each other.

(5) That if they are restrained in a contrary situation, they repel each other.

(6) That if a magnet is divided on a plane parallel to a line drawn through its poles, those parts of the segments which formerly were joined also flee each other.

(7) That if it is divided on a plane intersecting at right angles a line drawn through its poles, the two points previously contiguous form poles of different force, one in one segment and the other in the other segment.

(8) That, although in one magnet, there are only two poles, a South and a North, two similar poles are however also found in each of its fragments; so that its force, though seen to be different as far as its poles are concerned, is the same in any part as in the whole.

---

[92] Most of the properties listed here are mentioned in William Gilbert's *De Magnete* (London, 1600), which is considered to be the first scientific treatise on magnetism and which Descartes greatly admired. Descartes seems to have learned of most of the remaining properties from Mersenne; see note 107.

(9) That iron which is {touched by or} merely placed near to a magnet receives this force from it.

(10) That, according to the various ways in which the iron is placed near the magnet, it receives that force diversely.

(11) That an oblong piece of iron, placed near a magnet in any way whatever, always receives that force along its length.

(12) That a magnet loses nothing of its force, even if it communicates that force to the piece of iron.

(13) That this force is communicated to the piece of iron in a very short space of time indeed, but grows more and more stable in it with long duration of time, {if the iron remains next to the magnet in the same situation}.

(14) That the hardest steel receives greater force, and having received it maintains it more steadfastly, than does baser iron.

(15) That greater force is communicated to steel by a more perfect magnet, than by a less perfect one.

(16) That the Earth itself is also a magnet, and communicates some of its force to iron.

(17) That in the Earth, the largest magnet, this force seems less strong than in most other smaller ones.

(18) That needles which have been touched by a magnet turn their extremities toward {the poles of} the Earth, in the same way as a magnet does its poles.

(19) That these needles do not turn precisely toward the Earth's poles, but decline from them variously in various places.

(20) That this declination can change with time.

(21) That, according to certain men, there is no declination, or that perhaps there is not the same, or not such a great declination, in a magnet standing perpendicularly on one of its poles, as there is in that magnet when its poles are equidistant from the Earth.

(22) That the magnet attracts iron.

(23) That, when armed, a magnet supports much more iron than when unarmed.

(24) That its poles, although {otherwise} contrary, help each other to support the same piece of iron.

(25) That the rotation, in either direction, of little iron wheels suspended from a magnet, is not impeded by magnetic force.

(26) That the force of one magnet can be variously increased or decreased by affixing to it in various ways another magnet or a piece of iron.

(27) That a magnet, however strong, cannot draw a piece of iron away from the contact of a weaker magnet if it does not itself touch the iron.

(28) That, conversely, a weak magnet, or a tiny piece of iron, often separates a piece of iron contiguous to itself from a stronger magnet.

(29) That that pole of a magnet which we call its South pole supports more iron, in these Northern regions, than does that which we call its North pole.

(30) That iron filings arrange themselves in certain definite ways around one or more magnets.

(31) That a thin plate of iron, affixed to the pole of a magnet, deflects its force to attract and turn iron.

(32) That this same force is not impeded by the interposition of any other body.

(33) That a magnet which remains turned toward the Earth, or toward other nearby magnets, in a manner other than that in which it would spontaneously turn if nothing impeded its movement, loses its force with the passing of time.

(34) That, finally, this force is also diminished by rust, humidity, and moisture,[93] and removed by fire; but not by any other cause known to us.

146.   How the grooved particles flow through the pores of the Earth.

In order to understand the causes of these properties, let us consider the Earth AB,[94] of which A is the South pole, and B the North; and let us note that the grooved particles {represented by those little twisted bodies which have been drawn around it} coming from the Southern part E of the heaven are twisted in a completely different way from those coming from the Northern part F: as a result, it is absolutely impossible for those of one group to enter the other group's pores. Let us also note that the Southern particles proceed in a straight line from A to B through the middle of the Earth, and then return from B to A through the air surrounding the Earth; and at the same time, the Northern ones proceed from B to A through the middle of the Earth, and then return from A to B through the surrounding air: because the pores through which the grooved particles passed from one

[93] The term used here is 'situs', which usually means "situation" or "site"; however, it can also mean "moisture" or "mustiness". The French text omits the term both here and throughout Article 183.
[94] See Plate XXII, Fig. i.

side of the Earth to the other are such that the particles cannot return through them.

147.   That the grooved particles pass with more difficulty through the air, the water, and the exterior earth, than through the interior of the Earth.

Meanwhile, however, as new grooved particles are constantly arriving from parts E and F of the heaven, an equal quantity of others leave through the other parts G and H of the heaven, or are destroyed on the journey and lose their figures. This does not indeed happen when they are passing through the intermediate region of the Earth; because there they have pores hollowed out to their measurement, through which they flow very rapidly without any hindrance. But as they are returning through the air, the water, and the other bodies of the exterior earth, in which they have no pores of this kind; they move with much more difficulty and constantly encounter particles of the second and third element, by which they are crushed while trying to dislodge those particles.

148.   That they pass more easily through the magnet than through the other bodies of this exterior earth.

Now, if these grooved particles chance to come upon a magnet there, there is no doubt that they will pass through it much more easily than through the air or other bodies of the exterior earth; because, as was stated a bit earlier, they find pores in the magnet which are formed to their shape and arranged in the same way as the pores of the interior earth. At least [this will be true] when the magnet is so situated that the openings of its pores are turned toward those parts of the Earth from which those grooved particles which can freely enter those openings are approaching.

149.   Which are the poles of a magnet.

And, just as we call the central point of that part of the Earth in which are the openings of the pores which admit the grooved particles coming from the Southern part of the heaven its South pole; while the center of the other part (through which these particles leave and those coming from the North enter) is called its North pole: so too, we call the center of that part of the magnet with pores through which those particles [which originate in the

254 PART IV

Southern heaven] enter, its South pole; and the opposite point, its North
pole.[95] Nor do I care that others commonly call that pole which I call the
South, the North pole; {because they see that it naturally turns toward the
North, as I shall explain in a moment}. For the common people, who alone
have the right to render unsuitable names for things acceptable through
frequent use, are not accustomed to speak of the magnet; {and I am sure
that those who philosophize and desire to know the truth will not object to
my preferring reason to usage}.

150.   Why these poles turn toward the poles of the Earth.

However, when these poles of the magnet are not facing those parts of
the Earth from which come those grooved particles to which they can offer
free passage, then these grooved particles rush obliquely into the pores of
the magnet and drive it with that force which they have to continue their
movement along straight lines, until they have brought it back to its natural
situation {which is the most convenient for them}. And thus, whenever the
magnet is not restrained by some external force, [this action] causes its
South pole to be turned toward the North pole of the Earth, and its North
pole toward the South. This occurs because those grooved particles which
travel through the air {toward the magnet} from the North pole of the Earth
toward the South, first came from the Southern part of the heaven,
[passing] through the middle of the Earth [to its North pole]; and those
which are returning toward the North came from the North [part of the
heaven].

151.   Why the poles also incline toward the Earth's center at a
         definite angle.

The grooved particles also cause the magnet to incline one of its poles
more or less than the other toward the earth, depending on its location on
the earth. Thus, at the Equator, the South pole A of magnet L,[96] is directed
toward the North pole B of the Earth; and b, the North pole of the same

---

[95] This nomenclature was introduced by Gilbert, who realized that since the Earth itself was
magnetic, the North-seeking pole of a magnet was in fact its south magnetic pole.
[96] See Plate XXII, Fig. i.

magnet, is directed toward the South pole of the Earth; and neither pole of the magnet is depressed more than the other; because the grooved particles approach them from both sides with equal force. But at the North pole of the Earth, {the South} pole a of magnet N is entirely depressed, and pole b is raised to the perpendicular. In the intermediate places, however, magnet M raises its pole b more or less, and depresses pole a more or less, depending on its distance from pole B of the Earth. The cause of these variations is that the Southern grooved particles which are about to enter magnet N ascend from the interior parts of the Earth through pole B along straight lines; while the Northern particles from hemisphere DAC of the Earth come through the air on all sides toward the same magnet N, and must proceed no more indirectly to approach its higher part than to approach its lower: while the Southern particles which are about to enter magnet M, ascending from all that area of the Earth which is between B and M, have the force to depress its {Southern} pole a obliquely. And the Northern particles do not prevent this, since they approach pole b from the area AC of the Earth, [and they do this] as easily when pole b is raised as when it is depressed; {because in either case they must turn completely around in order to enter near pole b}.

152.   Why one magnet turns and inclines toward another, in the same way as toward the earth.

However, since these grooved particles flow through individual magnets in exactly the same way as through the Earth, they must turn two spherical magnets toward each other in the same way as they turn each {alone} toward the whole Earth. For it must be noted that the grooved particles always accumulate in much greater abundance around any magnet than in the other regions of the air: because of course, they have pores in the magnet through which they flow much more easily than through the surrounding air, which consequently keeps them next to the magnet {and thus they form a sort of vortex around a magnet just as they do around the Earth}. Similarly, because of the pores which they have in the interior of the Earth, there are more of them in the air and in the other bodies situated around the Earth, than in the heaven. And thus, as far as magnetic force is concerned, exactly the same things must be thought about one magnet in relation to another as about one in relation to the Earth, which itself can be said to be the largest magnet.

153.   Why two magnets approach each other, and what the sphere of
                  activity of each is.

Not only do two magnets turn toward each other until the North pole of
one is facing the South pole of the other; but in addition, {either while they
are turning or} after having thus turned, they approach each other until
they touch, if nothing hinders their movement. For it must be noted that
the grooved particles are very rapidly moved while they are situated in the
pores of the magnets; because they are carried along there by the impetus of
the first element, to which they belong. And when they emerge from there,
they encounter and drive the particles of other bodies which do not have as
much speed, since they belong to the second or third element. Thus, the
grooved particles which are passing through magnet O[97] acquire, from the
speed at which they are transported from A to B and from B to A, the force
to continue along straight lines toward R and S; until they there meet so
many particles of the second and third element that these turn them back
on both sides toward V. And the whole space RVS, {which contains the
vortex} through which they are thus scattered, is called the sphere of force
or activity of magnet O. This sphere obviously must be larger, the larger the
magnet is, and especially the longer it is along line AB; because the grooved
particles then proceed for a longer distance through the magnet and thus
acquire greater agitation. Similarly, the grooved particles which pass
through magnet P proceed in a straight line on both sides toward S and T;
and from there are turned back toward X, and drive all the air contained in
the sphere of the magnet's activity. But they do not thereby expel the air if
there is no place to which it can withdraw; and there is no such place when
the spheres of activity of these magnets are separated from each other. But
when they join to form one sphere [of activity]; then first, it is easier for the
grooved particles coming from O toward S to continue in a straight line to
P and replace those particles which previously returned from T via X
toward S and b, than it is for them to be turned back toward V and R
(toward which those coming from X proceed easily). And it is easier for
those coming from P to S to proceed to O than to be turned back toward X
(toward which those coming from V also proceed without difficulty). And
thus these grooved particles pass through these two magnets O and P, as if
they were a single one. Second, it is easier for the grooved particles
proceeding in a straight line from O to P, and from P to O, to expel the

---

[97] See Plate XXII, Fig. ii.

intermediate air from S toward R and T, into the place of magnets O and P (and thus to cause these magnets to approach each other until they touch each other at S), than for these particles to force their way through all this air from A to b, and from V to X; and these two paths become shorter when these two magnets approach each other, or, if one is restrained, at least when the other comes to it.

154.   Why they also sometimes recede from one another.

However, the like-named poles of two magnets do not thus approach each other, but on the contrary, if they are brought too close, recede instead. For the grooved particles emerging from that pole of one magnet which is facing [the like pole of] the other magnet are unable to enter that other magnet, and require some space between these two magnets through which they may pass in order that they may return to the other pole of the magnet from which they emerged. Specifically, since those coming out of O[98] through pole A cannot enter P through its pole a, they require some space between A and a through which they may pass toward V and B; and with the force which they have acquired by being moved from B to A, they drive magnet P. And similarly, those coming out of P drive magnet O; at least when their axes BA and ab are on the same straight line. However, when their axes are even very slightly inclined at an angle to one another, these magnets turn in the way explained a little earlier; or if their turning is impeded (but not their movement in a straight line), they once again recede from one another along a straight line. Thus, if magnet O is placed in a small boat floating on the water in such a way that its axis is always perpendicular, and if magnet P (whose South pole is facing the South pole of the other) is moved by a hand toward Y; this will cause magnet O to recede toward Z before it is touched by magnet P. For no matter in what direction the boat may turn, some space is always required between these two magnets in order that the grooved particles, coming out of them through poles A and a, can pass toward V and X.

155.   Why those parts of the segments of a magnet which were joined
            before the division also recede from each other.

And from these things, it is very easy to understand why it is that if a magnet is divided on a plane parallel to a line through its poles, and if one

[98] See Plate XXIII, Fig. 1.

segment is freely suspended {on a thread} above the magnet from which it
was cut off, that segment will spontaneously turn and assume a position
contrary to that which it previously had. So that, if parts A and a were
formerly joined,[99] and similarly B and b, afterwards b will turn toward A,
and a toward B. Of course, this happens because the Southern part of one
was previously joined to the Southern part of the other, and the Northern
to the Northern; but after their division, the grooved particles which have
emerged from the Southern part of the one, must enter through the
Northern part of the other; and those which have left through the
Northern, must enter through the Southern. {And by this means they cause
a, the South pole of the suspended segment, to turn toward B, the North
pole of the other, and b toward A}.

156.   Why two points, which were previously contiguous in one
            magnet, are poles of opposite force in its fragments.

It is also obvious why it is that if a magnet is divided on a plane at right
angles to a line drawn through its poles, the poles of those segments which
were contiguous to each other before the division (for instance poles b and
a)[100] are of opposite force: because the grooved particles which emerge
from one of these poles can enter only through the other.

157.   Why the force is the same in any part whatsoever of the magnet,
            as in the whole.[101]

It is no less obvious that the same [kind of] force is in any part whatever
of a magnet as in the whole: for this force is no different in the poles than in
the remaining parts, but only seems greater because the grooved particles
which have passed through the longest pores of the magnet emerge through
the poles, and they are located in the midst of all those coming from the
same direction. At least this is so in a spherical magnet, on the example of
which the poles in other magnets are considered to be in that place where
the greatest force appears. Nor is this force different in one pole than in the
other, except insofar as the grooved particles which have entered through
one, emerge through the other: but in any case there is no part of a magnet,

[99] See Plate XXIII, Fig. ii.
[100] See Plate XXIII, Fig. iii.
[101] See Article 145, property 8, for a clearer statement of the claim being made in this article.

however small, in which the grooved particles do not have an exit, if they also have an entrance; {and thus no part which does not have two poles}.

158.   Why a magnet communicates its force to a piece of iron which has been brought near to it.

Nor is it surprising that a piece of iron, brought near to a magnet, should acquire magnetic force from it. For iron has pores suited to receiving the grooved particles, {just as a magnet does}, and lacks nothing for the acquisition of this force, except that it has the tiny extremities of the small branches of which its scrapings consist protruding in [all] different directions in these pores; all of which must be bent in one and the same direction in those pores through which the grooved particles coming from the South can pass, and in the opposite direction in the other pores. However, when a magnet has been brought near, the grooved particles rush into the pores of the iron with great force and in great quantity, like a torrent, and thereby bend the extremities of the small branches; and consequently give to the iron all that it lacked in order to have magnetic force.

159.   Why iron receives this force diversely, according to the diverse ways in which it is placed near the magnet.

And further, according to the various parts of the magnet to which the piece of iron is applied, it receives this force variously. Thus, part R of piece of iron RST,[102] if applied to the North pole [B] of magnet P, will become the South pole of the piece of iron; because the Southern grooved particles will enter the iron through that part, and the Northern ones, which have been turned back through the air from pole A, will enter through part T. If the same part R lies above the equator of the magnet, and faces the magnet's North pole (as it does at C); it will again become the South pole of the piece of iron: but if it is turned around and faces the magnet's South pole (as at D); then it will lose the force of the South pole, and will become the North pole. Finally, if center part S of this piece of iron touches pole A of the magnet, the Northern grooved particles, having entered the piece of iron at S, will emerge from it at both ends through R and T; and thus the iron will receive the force of the South pole at both ends, and the force of the North pole at its center.

[102] See Plate XXIII, Fig. iv.

160.   Why an oblong piece of iron receives this force only along its
             length.

Yet it may be asked why these grooved particles, entering part S of the
piece of iron from pole A of the magnet, do not proceed in a straight line
toward E, but instead are turned back on both sides toward R and T; so
that this piece of iron receives the magnetic force along its length rather
than across its width. But the response is easy, because the grooved
particles find paths in the piece of iron which are much more open and easy
than those in the air, and they are consequently turned back by {the
resistance of} the air toward the piece of iron.

161.   Why a magnet loses nothing of its force although it com-
             municates this force to iron.

The reply is also easy if it is asked why a magnet loses nothing of its
force when it communicates that force to {a great quantity of} iron. For the
fact that the grooved particles which are emerging from a magnet enter a
piece of iron instead of any other body produces no change in the magnet
itself: except perhaps that, since they pass through the piece of iron more
freely than through other bodies, they also emerge from the magnet in
greater quantities when the piece of iron is attached to it; however, far from
being diminished by this, the magnet's force is increased instead.

162.   Why this force is very rapidly communicated to the piece of
             iron, but is stabilized in it over a period of time.

And this force is acquired by the piece of iron in the shortest time,
because the grooved particles flow very rapidly through it; but this force is
stabilized in the piece of iron over a long space of time, {if it is retained in the
same situation against a loadstone}; because the longer the extremities of
the small branches have remained bent in one direction, the more difficult it
is to turn them back in the opposite direction.

163.   Why steel is better suited to receiving magnetic force than baser
             iron.

And steel receives this force to a greater extent than does baser iron,
because it has more numerous and more perfect pores, suited to the

admittance of the grooved particles. And it retains this force more stably, because the extremities of the small branches protruding in these pores are less flexible.

164.   Why greater force is communicated to steel by a more perfect magnet than by a less perfect one.

And greater force is communicated to steel by a larger and more perfect magnet: both because the grooved particles rush into its pores with greater violence, and bend the extremities of the small branches protruding in those pores to a greater extent; and also because, inasmuch as more grooved particles simultaneously rush in there, they open up more such pores for themselves. For it must be noted that there are more such pores in steel, which of course consists solely of scrapings of iron, than in a magnet, in which there is much stony matter in which the scrapings of iron are embedded. Accordingly, since only a few grooved particles enter the iron from a weak magnet, they do not open all of its pores, but only a few, and moreover, only those which were closed by the most flexible extremities of the small branches. {And those grooved particles which come afterwards pass only through these open pores; so that the remaining pores are useless unless this iron is approached by a more perfect magnet which sends more grooved particles toward it}.

165.   Why the earth itself also imparts magnetic force to iron.

As a result, even base iron (in which of course the extremities of the tiny branches are extremely flexible) can receive some magnetic force in a very short time from the Earth itself; which is indeed the largest magnet, though a very weak one. Specifically, if the piece of iron is oblong and not yet endowed with any such force, and if one end of it is inclined toward the Earth, from that fact alone that end immediately acquires, in these Northern regions, the force of the South pole; and that end loses this force, and acquires exactly the opposite one, if it is raised, and the opposite one depressed.[103]

---

[103] The French text adds a description of an experiment to demonstrate this phenomenon, and states (incorrectly) that the acquired polarity of the depressed end will be retained if the iron is brought to a horizontal position.

166.   Why magnetic force is weaker in the Earth than in a small
            magnet.

But if it is asked why this force is weaker in the Earth, the greatest
magnet, than in other smaller ones, I reply that I do not think it weaker, but
rather that it is much stronger in that intermediate region of the Earth
which was stated above to be entirely penetrable by the grooved particles.
But I think that the grooved particles which emerge from that region return
primarily through that interior crust of the higher region of the Earth from
which metals originate, and in which there are also many pores suited to
admitting them. And therefore very few of those particles reach us. For I
judge that those pores in that interior crust, and also in the magnets and
scrapings of iron which are contained in the veins of this exterior region, are
twisted in the opposite way to the pores of the intermediate region: so that
the grooved parts which flow through this intermediate region from the
South to the North return from the North to the South through all the parts
of the higher region, but especially through its interior crust, and similarly
through the magnets and iron of the exterior [region]. And since most of the
grooved particles accumulate there, few remain to seek a path for
themselves through this air of ours and through the other surrounding
bodies which lack suitable pores. And if I correctly interpret these things, a
magnet cut from the earth, and freely placed in a boat upon the water, must
still turn that same surface which always previously faced the Northern
regions while it adhered in the earth toward the North: as Gilbert, the
principal investigator of magnetic force and the first discoverer of that
which is contained in the Earth, affirms he has proved.[104] Nor do I consider
it important that others think they have observed the opposite; for perhaps
they were deceived by the fact that, since the very part of the earth from
which they had taken care to separate the magnet, was itself a magnet; the
poles of the separated magnet turned toward it: for, as was stated earlier, a
fragment of one magnet turns [in a contrary way] toward another
fragment. {So, in order to perform this experiment correctly, after noticing
which sides of the magnet face the North and the South while it is joined to
the ore, one must remove it entirely from there and not place it near any
magnet other than the Earth}.

[104] See *De Magnete*, Book III, Chapter II.

167.   Why needles which have been touched by a magnet always have the poles of their force at their extremities.

Moreover, since this magnetic force is communicated to an oblong piece of iron only along its length: it is certain that a needle endowed with this force must always turn its extremities toward the same parts of the earth as a spherical magnet turns its poles {if it is at the same place on the Earth as the needle}; and that needles of this kind must always have the poles of their magnetic force precisely in their extremities.

168.   Why poles of magnetic force are not always directed precisely toward the poles of the Earth, but decline from them in various ways.

And because their extremities can more easily be distinguished from the remaining parts than can the poles of a magnet; with the help of such needles it has been noted that the poles of magnetic force do not everywhere precisely face the poles of the Earth, but decline from them variously in various places. The cause of this declination, as Gilbert noticed earlier, must be attributed solely to the inequalities which are in this surface of the earth.[105] For it is evident that many more scrapings of iron and many more magnets are found in some parts of the exterior earth than in others: with the result that the grooved particles which emerge from the interior earth flow in greater abundance toward certain places than toward others, and thus often turn aside from their courses. And because the turning of the poles of a magnet or of the extremities of a needle depends solely on the route of these particles; this turning must follow all the deviations of these particles. And it is possible to put this matter to the test in a magnet whose shape is not spherical: for if a slender needle is placed above various parts of that magnet, it will not always turn toward the magnet's poles in exactly the same way, but will often decline from them somewhat. Nor must it be thought that because the inequalities which are on the outermost surface of the earth are very tiny in comparison with the Earth's whole bulk, these inequalities are not the reason for this declination; for the inequalities should not be compared with that bulk, but with the needles and magnets in

[105] See *De Magnete*, Book IV, Chapters I and II.

which the declination occurs, and thus it will be apparent that they are sufficiently great.[106]

169.   How this declination can change sometimes with the passing of time {in a given area of the Earth}.[107]

There are some who say that this declination {is not only different in different places on the Earth, but} can also change with time at a given place; {so that the declination now observed in certain places does not coincide with that observed there in the last century}. This does not seem {to me} in any way strange, {considering that declination depends solely upon whether the quantity of iron and of loadstone happens to be greater or smaller on one side of these areas than on the other}; first, because men daily transfer iron from some parts of the earth to others; and second, because iron ore {deposits} in this exterior earth can become corrupted with time, and others can be created in other places {where there were none previously}, or sent up from the interior earth.

170.   Why this declination can be smaller, in a magnet standing vertically on one of its poles, than when its poles are equidistant from the Earth.

There are also those who say that this declination is nil in a spherical magnet standing perpendicularly on its South pole (in these Northern regions), or on its North pole in Southern regions; and that when this magnet has been placed in this position in a boat, it always turns one same point on its equator precisely toward the exact North, and the opposite point toward the South, {even when it is transported to various places}.

[106] The French text differs: "For although the inequalities on the Earth's surface are not very great in proportion to its whole bulk, . . . they are nevertheless sufficiently great . . . , in proportion to the various places on that surface, to cause the variation of the poles of the magnet which we observe."

[107] In 1639, Mersenne prepared a list of magnetic properties which was sent to Descartes and others; see Mersenne, *Correspondance*, VIII (Paris, 1963), 754–762. The list contained the statement that the declination at a given location was unvarying, an opinion held by Gilbert. During late 1639 and early 1640, Mersenne received several letters from John Pell, informing him that variations in declination had been observed in England by Gellibrand. Mersenne informed Descartes of this, and Descartes at that time conjectured that the variations might be caused by alterations in the Earth's topography. See A. & T., III, 46.

Whether this is true I have not yet ascertained by any experiment.[108] But I am easily persuaded that this declination is not in every respect the same, and perhaps is not as great, in a magnet thus placed as in a magnet whose poles are equidistant from the Earth. For, in this higher region of the Earth, the grooved particles do not only return from one pole to the other along lines equidistant from the Earth's center, but many also ascend from {or descend into} the Earth's interior parts, everywhere except near the equator: and the turning of a magnet standing vertically on its poles depends on these latter particles, whereas its declination depends mainly on the former.

171.   Why the magnet attracts iron.

Besides, the magnet attracts iron, or rather, a magnet and a piece of iron approach each other; for in fact there is no attraction there: rather, as soon as the iron is within the sphere of activity of the magnet, it borrows force from the magnet, and the grooved particles which emerge from both the magnet and the piece of iron expel the air between the two bodies:[109] as a result, the two approach each other in the same way as two magnets do. Admittedly, the iron moves more freely {toward the magnet} than the magnet {does toward the iron};[110] because iron is composed only of those scrapings in which the grooved particles have their pores, whereas the magnet is burdened with much stony matter.

172.   Why an armed magnet supports much more iron than an
              unarmed one.

However, many wonder that an armed magnet, or a magnet with a thin piece of iron attached {to one of its poles}, can support more iron than can

[108] Descartes learned of this claim in January, 1642, from a manuscript sent him by Constantine Huygens, who had received it from Mersenne. The author is unknown but was most probably Jacques Grandami, S.J., who claimed to have made the discovery in 1641. Descartes's letter to Huygens of January 31, states that he finds the claim very difficult to believe but that he does not think highly enough of his own speculations to have a spherical loadstone specially produced from him; see A. & T., III, 521–522. Descartes's explanation of this alleged property is essentially the same as one he suggested to Mersenne in 1643; see A. & T., III, 673.

[109] The French text differs and states that the air is expelled by the particles which pass from the magnet to the iron.

[110] This depends entirely on which of the two is the more massive, of course.

the magnet alone. But the reason for this can be discovered from the fact that even though an armed magnet supports more iron suspended from it, it does not, on that account, attract more to itself if it is even the slightest distance away from the iron; and that it does not even support more if some body, however thin, lies between the armed magnet and the iron. From this, it is evident that this greater force in the armed magnet arises solely from the difference of contact: because of course the pores of a thin piece of iron correspond perfectly to the pores of the iron suspended from it; and therefore, the grooved particles which pass through these pores from one piece of iron into the other expel all the intervening air, and cause the immediately contiguous surfaces of these pieces of iron to be very difficult to separate: and it has already been shown above that two bodies cannot be better united by any bond than by immediate contact. However, the pores of the magnet do not thus correspond to the pores of the iron (because of the stony matter which is in the magnet); and as a result, there must always remain some small quantity of space between the magnet and the iron, through which the grooved particles from the pores of one may reach the pores of the other. {And since these pores are not directly opposite, the grooved particles must flow somewhat obliquely between the two surfaces; and by retaining sufficient space for this, they prevent the surfaces from being completely contiguous}.

173.   Why its poles, although opposite, help each other to support iron.

Some also marvel that, although the poles of a magnet are seen to be of opposite force {as far as concerns their turning North and South}, they nonetheless help each other to support iron: so that if both poles are armed with thin sheets of iron, they can together support approximately twice as much iron as one alone. Thus, if AB is a magnet,[111] to whose poles are joined the thin iron plates CD and EF, which protrude on both sides in such a way that the piece of iron GH, which has been affixed to them, touches them on a fairly wide surface; this piece of iron GH can be approximately twice as heavy as if it were supported by only one of these thin metal sheets. But the reason for this is evident from the movement of the grooved particles which has already been explained. For although they are contrary in that those which enter through one pole cannot also enter through the

[111] See Plate XXIII, Fig. v.

other, this does not prevent them from uniting to support the iron; because those particles which emerge from the South pole A of the magnet, having been deflected by the steel plate CD, enter end b of the piece of iron, in which they create its North pole. And, flowing from b to the South pole a, they encounter the other steel plate FE, through which they ascend to B, the North pole of the magnet. And conversely, those particles which have emerged through B, return through the armature EF, the attached iron HG, and the other armature DC, to A. {And thus they join the iron to one of the armatures just as much as to the other}.

174.   Why the gyration of a small iron wheel is not impeded by the force of the magnet from which it has been suspended.[112]

However, this movement of the grooved particles through the magnet and through iron does not seem to be in harmony with the circular movement of small iron wheels which, after having been twirled like a spinning-top, rotate for longer when suspended [by their axle] from a magnet, than when they are far removed from the magnet and press upon the earth. And indeed, if the grooved particles were driven only by movements along straight lines, and if they found the individual pores of the iron (through which they must enter) exactly opposite the pores of the magnet (through which they are leaving); I should judge that these particles would have to halt the gyration of these wheels. But because these particles themselves constantly gyrate, some in one direction and the others in exactly the opposite direction, and must pass obliquely from the pores of the magnet into the pores of the iron; in whatever way a little wheel may be rotated, the particles enter its pores as easily as if it were at rest. And its motion is less impeded by contact with the magnet, when it is rotating thus suspended from that magnet {(because there always is some space between it and the wheel)}; than by contact with the Earth when it presses upon the Earth with its weight.

175.   How and why the force of one magnet increases or decreases the force of another.

The force of one magnet is increased or decreased in various ways by the approach of another magnet or of a piece of iron. But there is one general

[112] This property appears on Mersenne's list, see note 107.

rule in this matter: whenever these magnets are so situated that [each] one sends grooved particles into the other, they help each other; but, on the other hand, if [each] one causes fewer to go toward the other, they work against each other. This is because the more rapidly and abundantly these particles flow through each magnet, the greater is its force; and because more agitated and more numerous particles can be sent by one magnet or piece of iron into another magnet than can be sent by the air or any other body situated in its place, when it is absent. Thus, two magnets not only help each other to support iron when the South pole of one magnet is joined to the North pole of the other and the iron is suspended from their other poles, but also when they are disunited, and the iron is placed between the two. For example, magnet C[113] is helped by magnet F to retain the piece of iron DE which is joined to C. And conversely, magnet F is helped by magnet C to support extremity E of this piece of iron in the air: for E can be so heavy that it could not thus be supported by F alone, if the other extremity D were resting upon some body other than magnet C.

176.   Why a magnet, however strong, cannot attract to itself a piece of iron which is contiguous to a weaker magnet.

Nevertheless, a certain force of magnet F, namely that which it has to attract to itself the piece of iron DE, is impeded by magnet C. For it must be noted that as long as this iron touches magnet C, it cannot be attracted by magnet F which it is not touching; even if we suppose F to be much more powerful than C. The reason for this is that the grooved particles pass through these two magnets and this piece of iron as if through a single magnet, in the manner explained above, and have approximately equal force in the whole space which is between C and F. And, on that account, they cannot draw toward F piece of iron DE, which is not only bound to magnet C by magnetic force, but also by contact.

177.   Why a weak magnet, or a piece of iron, can draw away from a stronger magnet a piece of iron contiguous to itself.

And from this it is obvious why often a weak magnet, or a slender piece of iron, draws another piece of iron away from a stronger magnet {to which it is attached}. For it must be noted that this never occurs except when the

[113] See Plate XXIII, Fig. vi.

weaker magnet is touching that piece of iron which it draws away from the stronger magnet. For when the extremities of an oblong piece of iron are touched by dissimilar poles of two magnets, and when these two magnets are then moved away from each other, the piece of iron between them does not always adhere to the weaker magnet, or even always to the stronger, but sometimes to one and sometimes to the other. And I think {that this shows that} there is no reason why it should adhere to one rather than to the other, except that it touches the one to which it adheres on a greater surface-area.

178. Why, in these Northern regions, the South pole of a magnet is more powerful than the North pole.[114]

However, from the fact that magnet F aids magnet C to support piece of iron DE, it is obvious why that pole of the magnet, which we call its South pole, {seems to have more power and} supports more iron than the other, in these Northern regions. For its South pole is helped by the Earth, the largest magnet, {when that pole is turned toward the North pole of the Earth}, in exactly the same way as magnet C by magnet F; whereas the other pole, because of its inappropriate situation, is impeded by the Earth, {when that pole is turned [downward] toward the Earth in this hemisphere}.

179. Concerning those things which can be observed in iron filings scattered around a magnet.

If we consider, a little more carefully, the way that iron filings arrange themselves around a magnet, we shall, with their help, notice many things which will confirm what has already been said. For, first of all, it can be noticed that they do not accumulate at random, but that they incline toward each other, forming, as it were, certain small conduits through which the grooved particles flow more freely than through the air, and which accordingly indicate the paths of these particles. In order that these paths may be clearly perceived by the eye, let some of these iron filings be sprinkled over a plane containing an aperture into which a spherical magnet has been inserted in such a way that it touches the plane with its

---

[114] This property is mentioned by Gilbert (Book II, Chapter XXXIV). The claim that the north pole of a loadstone "attracts more iron" is included in a very brief set of notes on Kircher's *Magnes sive De Arte Magnetica* (Rome, 1641) which Descartes made early in 1643. The additions to the French text bring this article into much closer agreement with Gilbert, who specifically states that if each pole is turned upward, the north will be the more powerful.

poles on both sides (in the manner in which Astronomers' globes are usually inserted into the circle of the Horizon so that they may represent the right sphere[115]); and the iron filings scattered there will arrange themselves into small conduits which will show the curved paths of the grooved particles around the magnet; or even those around the globe of the Earth which were previously described. Then, if another magnet is introduced in the same way into this plane next to the first one,[116] and if the South pole of one is facing the North pole of the other; the filings which have been scattered around will also show how the grooved particles are moved through these two magnets as if through a single one. For the conduits which will extend between the poles which are facing each other will be entirely straight; while the other small conduits (which will reach from one of the poles which are turned away from each other to the other) will be curved around the magnet: as are here lines BRVXTa.[117] Moreover, when some iron filings are suspended from the pole of one magnet, the South for example; if the South pole of a second magnet, which has been placed beneath the first, is turned toward these filings, {which will be hanging down vertically}, and is gradually moved closer to them, it is possible to observe how the small conduits formed by these filings at first draw themselves back upward, and bend:[118] because of course those grooved particles which flow through the conduits are repulsed by others coming from the lower magnet. And then, if this lower magnet is much more powerful than the upper one, these conduits are disintegrated, and the filings fall down onto the lower magnet: because of course the grooved particles ascending from this lower magnet strike violently against the individual filings; and since they are unable to enter them except through those surfaces by which the filings are adhering to the upper magnet, they detach the filings from this upper magnet. On the other hand, if the North pole of the lower magnet is turned to face the South pole of the upper, to which the iron filings are adhering, these iron filings direct their conduits in a straight line toward the lower magnet, and extend them as far as possible:[118] because of course these conduits offer passage in both

[115] The right sphere is the celestial sphere as viewed from the Earth's equator, i.e., with the celestial poles on the horizon.
[116] See Plate XXII, Fig. ii.
[117] It is not known how Descartes learned of these effects; they are not in any of the sources he is known to have used nor does any mention of them appear in his correspondence. In fact, this seems to be the earliest detailed description of these phenomena.
[118] These effects are included in Mersenne's list; see note 107.

directions to the grooved particles crossing from one magnet into the other. But the filings are not thereby separated from the upper magnet, unless they have previously touched the lower one; because of the force of contact which we discussed a little earlier. And because of this same force, if the iron filings adhering to any magnet, however powerful, are touched by another weaker magnet or merely by some iron rod, part of them will leave the stronger magnet, and will follow the weaker one or the iron rod: specifically, those [filings] which touch the latter on a larger surface than they do the former. For, inasmuch as these tiny surfaces are diverse and uneven, it often happens that they join certain of the filings more firmly to one magnet or piece of steel than to the other.

180.   Why an iron plate joined to the pole of a magnet hinders its power to attract or turn iron.

An iron plate which has been placed against the pole of a magnet greatly increases its power to support iron, as was stated above; but it impedes the force of the same [magnet] to attract or turn iron toward itself. Thus, sheet DCD,[119] attached to the pole of magnet AB, prevents it from attracting or turning toward itself needle EF. For we notice that the grooved particles which would proceed from B toward EF, were it not for this plate, are now deflected at the plate from point C toward the extremities DD; because these particles flow more freely through the plate than through the air, and thus scarcely any reach needle EF. Similarly, we stated above that few grooved particles from the intermediate region of the Earth reach us, because most of them return through the interior crust of the higher region of the Earth, from one pole to the other; as a result, only a weak magnetic force of the entire Earth is felt here among us.

181.   Why the interposition of no other body has the same effect.

However, apart from iron or a magnet, it is not possible to put any other body in place of plate CD which will impede magnet AB from exerting its force on needle EF. For, in this exterior earth, we have no body, however solid and hard, in which there are not very many pores: not indeed formed to the measurement of the grooved particles, but much larger, since they also admit the globules of the second element. And these grooved particles

[119] See Plate XXIII, Fig. vii.

can consequently pass through these pores as freely as through the air, in which they also encounter these globules of the second element.

182.   Why an inappropriate position of the magnet gradually decreases its force.

If a piece of iron or a magnet is kept for a long time turned toward the Earth or other nearby magnets in a manner other than that in which it would turn of its own accord (if nothing impeded its movement); by that fact alone, it gradually loses its force: because then the grooved particles, arriving from the Earth or from the other nearby magnets, encounter its pores obliquely or inappropriately, and gradually change and damage their figures.

183.   Why rust, humidity, and moisture also diminish this force, and why intense fire completely removes it.

Finally, magnetic force is much diminished by humidity, rust, and moisture; and completely removed by a hot fire. For rust, springing up out of the scrapings of the iron, closes the entrances to the pores; and humidity of the air and moisture do the same thing, because they are the origins of rust. However, the agitation of fire throws the position of these scrapings into complete disorder. And I do not think that up to this time, anything has truly and for a certainty been observed concerning the magnet, the reason for which is not easily understood from those things which I have explained.

184.   Concerning the force of attraction in amber, wax, resin, and other similar things.

However, apropos of the magnet which attracts iron, something must be added here concerning amber, jet, wax, resin, glass, and similar things; which all also attract tiny bodies. For although it is not my intention to explain any particular things except insofar as they are required to confirm the more general things with which I have been concerned; and although I cannot examine this force in jet or in amber without first deducing from various observations many of their other properties, and thus investigating their innermost nature: nevertheless, because the same force is also in glass (which I was obliged to discuss a little earlier in order to demonstrate the effects of fire), if I did not explain this force, perhaps the other things which I have written about glass could be called into doubt. Especially since some men, seeing that this force occurs in amber, in wax, in resin, and in practically all oily substances, will perhaps think it consists in the fact that certain slender and branching particles of these bodies, having been moved by friction (for friction is usually required to arouse this force), scatter themselves through the nearby air, and, adhering to one another, immediately return and bring with them the tiny bodies which they strike on their way. Just as we see that a {very sticky} drop of liquified fats of this kind, suspended from a rod, can be shaken by slight movement in such a way that one part of the drop still adheres to the rod, while another part descends for some distance and immediately returns {of its own accord toward the rest of the drop} and also brings with it the tiny straws or other minute bodies which it has encountered. For no such thing can be imagined in glass, at least if its nature is as we described it above; and therefore another cause of this attraction in it must be indicated.

185.   What the cause of this attraction in glass is.

Of course, from the manner in which it has been indicated that glass is produced, it is easily inferred that besides those fairly large interstices through which the globules of the second element can pass in all directions, there are also many fairly long fissures between its particles; and, being too narrow to admit these globules, these fissures offer passage only to the matter of the first element. And it must be thought that this matter of the first element (accustomed to assuming the figures of all the pores which it enters), by passing through these fissures, is formed into certain thin, wide, and fairly long ribbons, which remain within the glass, or certainly do not stray far from it since they do not find similar fissures in the surrounding air. And, twisting around the particles of the glass with a certain circular movement, these ribbons flow from some of the fissures into others. For even though the matter of the first element is extremely fluid, still, because it is composed of tiny particles which are unequally agitated (as I explained in Articles 87 and 88 of Part III), it is perfectly in accordance with reason for us to believe that many of its most agitated particles constantly depart from the glass into the air, and that others return in their place from the air into the glass. But since those which return are not all equally agitated, those which have the least agitation are driven into the fissures, to which no pores in the air correspond; and there they adhere to one another, and form these ribbons. Accordingly, with the passing of time, these ribbons acquire stable figures which they cannot easily change. As a result, if glass is rubbed vigorously enough, so that it grows slightly warm; these ribbons will be shaken out by this movement, and scatter through the nearby air and also enter the pores of other neighboring bodies. But because they do not so easily find paths there; they immediately return to the glass, and bring with them those tinier bodies in whose pores they are caught.

186.   That the same cause of this attraction is also seen in the remaining things.

Moreover, what we have here noted concerning glass must also be believed of most other bodies: to wit, that certain interstices are found between their particles which are too narrow to admit the globules of the second element, and accept only the matter of the first. And since these interstices are larger than those in the surrounding air which are similarly open only to this matter of the first element; they are filled by the less agitated of its tiny particles. These adhere to one another and form particles which, of course, have diverse figures (because of the diversity of these interstices), but which are for the most part thin, wide, and fairly long, like ribbons. Thus, they can be constantly moved by twisting around the particles of the bodies in which they are. For, inasmuch as the interstices from which these ribbons acquire their figure must be extremely narrow in order not to admit the globules of the second element; unless they were oblong, like fissures, they could scarcely be larger than those which are between the particles of air but which are not occupied by the globules of the second element. On that account, although I do not deny that the other previously indicated cause[120] of attraction could perhaps occur in some bodies, yet because it is not so general, and because attraction is observed in very many bodies; I do not think that any cause other than that found in glass is to be sought in these bodies, or at least in most of them.

187.   That, from what has been said, it is understood what the causes of all the remaining wonderful effects usually attributed to occult qualities can be.

In addition, I wish it to be observed here that these particles, composed of the matter of the first element in the pores of terrestrial bodies,

[120] That is, the cause described in Article 184.

can not only be the causes of various attractions, such as are in amber and in the magnet, but also of innumerable other admirable effects. For those particles which are formed in [the interstices of] each body have something individual in their figure, by which they differ from all those formed in other bodies. And since they retain the greatest agitation of the first element, of which they are parts; it can occur, from the slightest of causes, that they either do not stray outside the body in which they are (but only rush this way and that within its pores); or, on the other hand, depart from that body very rapidly, and, passing through all other [kinds of] terrestrial bodies, reach places as distant as you please in the shortest time, and, finding matter suited to receiving their action, produce some rare effects there.[121] And certainly, if anyone will consider how marvelous are the properties of the magnet and of fire, and how different they are from those which we commonly observe in other bodies; how huge and forceful a fire can be ignited from the tiniest spark in an instant; to what an immense distance the fixed stars send their light on every side; and the remaining things whose causes (sufficiently obvious, in my judgment) I have deduced in this piece of writing from principles known and accepted by all (namely, from the figure, magnitude, situation, and motion of particles of matter): he will be easily persuaded that there are, in rocks or plants, no forces so secret, no marvels of sympathy or antipathy so astounding, and finally, no effects in all of nature which are properly attributed to purely physical causes or causes lacking in mind and thought; the reasons for which cannot be deduced from these same principles. Consequently, it is unnecessary to add anything else to them.

188.   Concerning those things which must be borrowed from [my projected] treatises on animals and on man, for an understanding of material things.

I should not add more to this fourth part of *Principles of Philosophy*, if (as I previously intended) I were still going to write two other parts; that is, a fifth concerning living things, or animals and plants; and a sixth concerning man. But because all the things which I would wish to discuss in those parts have not yet been perfectly examined by me, and because I do not know whether I shall ever have sufficient leisure {or experiments} to

---

[121] As examples, the French text mentions dreams, premonitions, and telepathy; which would not seem to be physical events.

complete them; instead of further delaying these earlier parts, or allowing anything which I might have reserved for the others to be lacking in them, I shall add here some few things concerning the objects of the senses. For up to now; I have described this Earth, and indeed this whole visible world, as a machine, considering nothing in it except figures and motions; yet our senses show us many other things, namely, colors, odors, sounds, and similar things. And if I remained completely silent about these, I should seem to have omitted the principal part of the explanation of natural things.

189.   What sensation is, and how it occurs.

And so it must be known that even though the human soul directs[122] the whole body, it nevertheless has its principal seat in the brain; there alone it not only understands and imagines, but also senses. And it senses by means of the nerves which are extended like threads from the brain to all the remaining members. These nerves are connected to these members in such a way that scarcely any part of the human body can be touched without some of the extremities of the nerves distributed through it being moved simply by that fact, and without this movement being transmitted to the other extremities of those nerves, which are accumulated in the brain around the seat of the soul; as I explained at sufficient length in Chapter 4 of the *Dioptrics*. However, the movements which are thus excited in the brain by the nerves affect the soul or mind (which is closely joined to the brain), in diverse ways, according to the diversity of these movements. And these diverse states or thoughts of the mind, following immediately from these movements, are called the perceptions of the senses, or, as we commonly say, sensations.

190.   Concerning the distinction of the senses: and first, concerning the internal ones, that is, the {passions or} states of the rational soul,[123] and the natural appetites.

The diversities of these feelings depend first, on the diversity of the nerves themselves, and second, on the diversity of the movements which occur in individual nerves. However, each individual nerve does not create an

[122] Latin: '*informo*'; 'instruct'. In Scholastic usage, the term meant "give substantial form to"; cf. Article 198.
[123] Latin: '*animus*'; 'the soul as the seat of thought or feeling'.

individual feeling different from all other feelings, for only seven principal distinctions can be noted in them; two of which concern internal feelings, and the other five external ones. For of course the nerves which extend to the stomach, the esophagus, the throat, and the other interior parts intended to satisfy natural needs, form one of the internal senses, which is called natural appetite. However, the small nerves which extend to the heart and to the praecordia,[124] although extremely tiny, create the other internal sense; in which consist all the stirrings, or passions, and all the states of the rational soul, such as happiness, sorrow, love, hatred, and similar things. Thus, for example, if the blood is properly moderated, it dilates in the heart easily and more than usual, and relaxes and moves the small nerves distributed around the orifices of the heart in such a way that another movement in the brain follows from this; this movement affects the mind with a certain natural feeling of cheerfulness: and any other causes which move these small nerves in the same way give this same feeling of happiness. Thus, imagining the enjoyment of something pleasant does not itself contain the feeling of happiness; but it sends spirits from the brain to the muscles in which those nerves are implanted, and with the help of these muscles, the orifices of the heart are expanded, and its small nerves are moved by that movement from which that feeling must follow. Thus, when agreeable news has been heard, the mind first judges of it, and rejoices with that intellectual joy which occurs without any excitation of the body, and which, on that account, the Stoics said could be suited to a wise man. Then when the news is imagined, spirits flow from the brain to the muscles of the praecordia, and there move small nerves with whose help these spirits excite another movement in the brain, which affects the mind with a feeling of animal happiness. Similarly, blood which is too thick flows scantily into the ventricles of the heart and dilates insufficiently there; this creates a certain other movement in these small nerves of the praecordia which is communicated to the brain and thereby places a feeling of sadness in the mind; although the mind itself may perhaps not know why it is saddened: and many other causes achieve this same effect. Moreover, other movements of these small nerves cause other states; for example, love, hatred, fear, anger, etc., insofar as these are only states, or passions, of the rational soul. That is, insofar as they are certain confused thoughts, which the mind does not have from its own nature,[125] but, rather, because something is

---

[124] The praecordia is the space beneath the heart, once thought to be the seat of the feelings.
[125] Literally, "of itself alone".

being experienced by the body to which the mind is closely joined. For the
distinct thoughts which we have concerning those things which ought to be
cherished, or chosen, or avoided, etc., are of a completely different kind
from these states. There is no other reason for natural appetites, such as
hunger, thirst, etc., which {are feelings excited in the soul and} depend on
the nerves of the stomach, the throat, etc., and which are completely
different from the will to eat, drink, etc.; but because, most of the time, this
will or striving[126] accompanies them, these [sensations] are called
appetites.

191.   Concerning the exterior senses: and first, concerning the sense
              of touch.

As for the exterior senses, five are commonly counted, because of five
diverse kinds of objects moving the nerves which serve these senses, and
because of an equal number of kinds of confused thoughts, which are
excited in the soul by these movements. For, first, the nerves ending in the
skin of the entire body, by the intermediary of that skin can be touched by
any terrestrial bodies whatever, and moved by all their properties;[127] in
one way by their hardness, in another by their weight, in another by their
heat, in another by their humidity, etc. And in however many diverse ways
they are either moved or impeded from their normal movement, these
nerves excite an equal number of diverse feelings in our mind; after which
[feelings] an equal number of tactile qualities are named. {And the names of
these qualities, such as 'hardness' . . . etc., mean nothing except that there
is, in these bodies, whatever is required to cause our nerves to excite feelings
of hardness, weight, heat, etc., in our soul.} And moreover, when these
nerves are agitated more vigorously than usual, and yet in such a way that
no hurt to the body ensues from this; this causes a feeling of titillation,
naturally pleasing to the mind (because it proves, to the mind, the strength
of the body to which it is closely joined). If, however, {this action is even

---

[126] The term used here is '*appetitio*'; the term which has been translated as 'appetite' is
'*appetitus*': the French text ignores the distinction. In a letter written October 6, 1645,
Descartes states, "It is true that we almost never have any thoughts which do not depend on
several of the causes I have just distinguished; but we give these thoughts the name of the
principal cause or of the cause with which we are mainly concerned. This causes many . . . to
confuse. . . the feelings of thirst or hunger with the desires to drink or eat, which are [properly
called] passions.": A. & T., IV, 311–312.
[127] Literally: 'and moved by them entire'.

slightly stronger so that} some injury results from this, it causes a feeling of pain. And from this it is obvious why physical pleasure and pain are so similar to each other in the object {causing them}; although they are contrary in feeling.

192.   Concerning taste.

Then other nerves, distributed throughout the tongue and the parts adjacent to it, are moved by particles of the same bodies which are separated from one another and floating in the mouth along with the saliva. And these nerves are variously moved according to the diverse shapes {or movements} of the particles; thereby creating the sensations of diverse tastes.

193.   Concerning the sense of smell.

Thirdly, two nerves or appendages of the brain (which do not protrude beyond the skull) are also moved by particles of the same bodies which are separate and flying in the air. Not indeed by any sort of particles whatever; but only by those which are sufficiently subtle and at the same time sufficiently animated to penetrate to these nerves through the pores of the spongy {ethmoid} bone, after having been drawn into the nostrils. And the diverse movements of these nerves create the sensations of diverse odors.

194.   Concerning hearing.

Fourthly, two other nerves, hidden in the innermost cavities of the ear, capture the tremulous and impetuous movements of all the surrounding air. Because the air, striking upon the small membrane of the tympanum, immediately shakes the attached chain of three small bones, to which these nerves adhere; and the sensations of diverse sounds result from the diversity of these movements.

195.   Concerning sight.

Finally, the extremities of the optic nerves, which form the membrane in the eyes called the retina, are not moved there by the air or by any terrestrial bodies, but solely by the globules of the second element;[128] whence there occurs a sensation of light and color; as I have already sufficiently explained in the *Dioptrics* and in the *Meteorology*.[129]

196.   That the soul does not feel except insofar as it is in the brain.

Moreover, it is clearly proved that the soul feels what happens to the body (by means of the nerves in the body's individual members), not insofar as the soul is in those individual members; but only insofar as it is in the brain. First, [this is proved] by the fact that various illnesses which affect only the brain remove or disturb all feeling; as sleep itself (which is in the brain alone) every day deprives us for the most part of the ability to feel, which awakening afterward restores to us. Next, from the fact that if the brain is in a sound state, and the paths of the nerves which extend to the brain from the external members are merely obstructed; by that single fact, the feeling of these members also perishes. And finally, from the fact that pain is sometimes felt as if it were in certain members, though there is no cause of pain in these members, but rather in other members, through which the nerves which stretch from these [painful] members to the brain pass. This last point can be shown by innumerable experiences, but it will suffice to indicate one here. When the eyes of a certain girl whose hand was infected by a serious disease were blindfolded whenever the Surgeon approached (lest she might be disturbed by the apparatus of treatment); and when, after some days, her arm had been amputated up to the elbow, on account of the gangrene spreading through it; and when cloths had been substituted for the amputated part, in order that she might be completely ignorant of having been deprived of it: she would sometimes complain that she felt various pains in the hand which had been removed, now in one finger, now in another. This clearly could not happen for any reason other than that the nerves which previously descended from the brain to the hand, and were then terminated in the arm next to the elbow, were moved there in the same way as they must previously have been in the hand when the feeling of this or that painful finger was imprinted upon the soul residing in the brain. {And this clearly shows that a pain in the hand is not felt insofar as the soul is in the hand, but only insofar as it is in the brain}.

[128] The French text here adds the claim that the globules pass through the transparent parts of the eye; an assumption which Descartes rejects as unnecessary in the *Dioptrics*.
[129] See *Dioptrics*, Discourse VI; and *Meteorology*, Discourse VIII.

197.   That the mind is of such a nature that various feelings can be excited in it solely by the movement of a body.

Next, it is proved that the nature of our mind is such that, simply from the fact that certain movements occur in a body, it can be driven to all sorts of thoughts, which convey no image of these movements; and especially to those confused thoughts which are called feelings or sensations. For we see that either spoken or even written words can excite any thoughts and stirrings whatever in our mind. On the same sheet of paper, with the same quill and ink, if the end of the quill is merely guided over the paper in a certain way; it will produce letters which will excite thoughts of combats, tempests, and furies, and states of indignation and sadness in the minds of readers. If, however, the quill is moved in another almost identical manner, it will cause very different thoughts, of calm weather, peace, and pleasantness, and exactly opposite states of love and happiness. It will perhaps be replied that writing or speech does not excite states and images of things different from itself directly in the mind, but only diverse understandings; on the occasion of which the soul itself, {which understands the meaning of these words}, forms in itself images of various things. But what will be said of the feelings of pain and titillation? A sword, applied to our body, cuts it: from this alone pain is produced, {without thereby indicating to us what the movement or figure of the sword is. The idea of} this pain is obviously as different from the local motion of the sword or of the body which is cut, as is {the idea of} color, sound, odor, or flavor. And since we clearly see that the feeling of pain is excited in us solely by the fact that some parts of our body are locally moved by contact with some other body, we can on that account conclude that our mind is of such a nature that, from some other local motions, it can have the experiences of all the other feelings [and sensations].

198.   That we perceive by our senses nothing in external objects except their figures, sizes, and movements.

Furthermore, we do not perceive any differences between nerves from which it might be permissible to judge that there is anything which reaches the brain (from the organs of the external senses) through some nerves but not through others; or that anything other than local movement of these nerves reaches it at all.[130] And we see that this local movement not only

[130] In the *Dioptrics*, Descartes implies that differences between sensations are due to the different locations of the nerve-endings in the brain and to the different sorts of movements transmitted by the nerves; see Discourses IV and VI.

produces the feeling of titillation or pain, but also that of light and sounds. For if anyone is struck in the eye in such a way that the vibration of the blow reaches the retina, that alone will cause him to see very many sparks of flashing light which will not be outside the eye. And if someone stops up his ear with a finger, he will hear a certain tremulous murmur, which will result solely from the movement of the air trapped in the ear. Finally, we often notice that heat and other perceptible qualities, insofar as they are in objects, and also the forms of purely material things (as for example, the form of fire), arise from the local movement of certain bodies, and that these then themselves cause other local movements in other bodies. And we very well comprehend how the various sizes, figures, and movements of the particles of one body produce various local movements in another body. However, we cannot in any way comprehend how the same things (that is, size, figure, and movement) can produce something else of an entirely different nature from themselves, such as those substantial forms and real qualities which many {Philosophers} suppose to be in things; nor indeed how, subsequently, these qualities or forms can have the force to excite local movement in other bodies. Since this is so, and since we know it to be the nature of our soul that diverse local movements suffice to provoke in it all feelings; and since we know by experience that those various feelings are in fact aroused in it, and do not perceive that anything other than movements of this kind travels to the brain from the organs of the external senses: it must certainly be concluded regarding those things which, in external objects, we call by the names of light, color, odor, taste, sound, heat, cold, and of other tactile qualities, or else [by the names] of substantial forms; that we are not aware of their being anything other than various arrangements {of the size, figure, and motions of the parts} of these objects which make it possible for our nerves to move in various ways, {and to excite in our soul all the various feelings which they produce there}.

199.   That no phenomena of nature have been omitted in this
              treatise.

And thus, by simple enumeration, it is concluded that no phenomena of nature have been omitted by me in this treatise. For nothing is to be numbered among the phenomena of nature, except what is perceived by the senses. However, apart from size, figure, and motion, [the varieties of] which I have explained as they are in each body, nothing located outside us is observed except light, color, odor, taste, sound, and tactile qualities;

which I have now demonstrated are nothing in the objects other than, or at least are perceived by us as nothing other than, certain dispositions of size, figure, and motion {of bodies. Thus,...there is nothing visible or perceptible in this world that I have not explained}.

200.   That I have used no principles in this treatise which are not accepted by all; and that this Philosophy is not new but extremely ancient and commonplace.

However, I should also like it to be noted that I have here attempted to explain the entire nature of material things in such a way that I have used, for this purpose, absolutely no principle which was not accepted by Aristotle and by all other Philosophers of all periods: so that this Philosophy is not new, but the oldest and most commonplace of all. For of course I have considered the figures, motions, and sizes of bodies, and have examined, according to the laws of Mechanics (which are confirmed by certain and daily experiences), what ought to follow from the collision of these bodies. Yet who ever doubted that bodies are moved, and are moved variously according to their various sizes and figures; or that as a result of the collision of these bodies, the larger ones are divided into many smaller ones, and change their figures? We do not observe this through only one sense, but through several: through sight, touch, and hearing [sic]; and we also {very} distinctly imagine and {clearly} understand this. This cannot be said of the remaining qualities {perceived by our senses}, like colors, sounds, and the rest, which are perceived not by means of several senses, but only by means of individual ones: for their images in our minds are always confused, and we do not know what they may be.[131]

201.   That imperceptible[132] particles of bodies exist.

I also consider, in individual bodies, many particles which are not perceived by sense: which may not be approved by those who take their senses as the measure of the things they can know. Yet, if only he considers what is added each hour to those bodies which are gradually being increased,

[131] The final sentence in the French text is: "This cannot be said of the remaining things which our senses perceive,...: for each of these things touches only one of our senses, and only impresses on our imagination a very confused idea of itself, and finally does not make known to our understanding what it is [in itself]."
[132] See Part I, note 1.

or what is removed from those which are being decreased; who can doubt
that there are many bodies so tiny that we do not perceive them by our
senses? A tree grows each day, but it cannot be understood to be made larger
than it previously was unless it is likewise understood that some body is
added to it. But who ever perceived by the senses those small bodies which in
a single day are added to a growing tree? And, at least, those who know that
quantity is indefinitely divisible will have to acknowledge that it must be
possible for its parts to be made so small as to be unperceivable to any sense.
And we certainly should not be surprised that we are unable to feel extremely
minute bodies; since our nerves themselves (which must be moved by objects
in order to create sensation) are not extremely tiny, but are like thin cords,
composed of many particles smaller than themselves; and thus they cannot
be moved by the most minute bodies. Nor do I think that anyone who is
using his reason will be prepared to deny that it is far better to judge of things
which occur in tiny bodies (which escape our senses solely because of their
smallness) on the model of those which our senses perceive occurring in large
bodies, than it is to devise I know not what new things, having no similarity
with those things which are observed, in order to give an account of those
things [in tiny bodies]. {E.g., prime matter, substantial forms, and all that
great array of qualities which many are accustomed to assuming; each of
which is more difficult to know than the things men claim to explain by their
means.}

202.   That the Philosophy of Democritus differs as much from ours
              as from the generally accepted one.

    Yet Democritus also imagined certain small bodies, having various
figures, sizes, and movements, from the accumulation and collision of
which all perceptible bodies arose; however his method of philosophizing is
commonly rejected by all. But no one has ever rejected it on account of the
fact that it considers certain bodies which are so minute that they escape the
senses, and which are said to have various sizes, figures, and movements;
because no one can doubt that there are indeed many such bodies, as has
just been shown. But it has been rejected, first, because it supposed those
small bodies to be indivisible, for which reason I also reject it. Second,
because it imagined that there was a void around these bodies, which I
show cannot be. Third, because it attributed weight to these bodies,
whereas I understand that there is no weight in any body considered in
isolation, but only insofar as that body depends on the situation and

movement of other bodies, and relates to them. And finally, because it did not show how individual things resulted solely from the encounters of small bodies, or if it showed this about some things, not all of its reasons were consistent with one another: at least as far as it is permissible to judge from those of his opinions which have been recorded. However, I leave it to others to judge whether those things which I have so far written about Philosophy are sufficiently coherent, {and whether sufficient things can be deduced from them. And because consideration of figures, sizes, and motions was accepted by Aristotle, as well as by all the others; and because I reject everything else which Democritus assumed (as I on the whole reject everything assumed by the others); it is evident that this way of philosophizing has no more affinity with that of Democritus than with that of all the other sects}.

203. How we know the figures and movements of imperceptible particles.

But I attribute determined figures, and sizes, and movements to the imperceptible particles of bodies, as if I had seen them; and yet I acknowledge that they are imperceptible. And on that account, some readers may perhaps ask how I therefore know what they are like. To which I reply: that I first generally considered, from the simplest and best known principles (the knowledge of which is imparted to our minds by nature), what the principal differences in the sizes, figures, and situations of bodies which are imperceptible solely on account of their smallness could be, and what perceptible effects would follow from their various encounters. And next, when I noticed some similar effects in perceptible things, I judged that these things had been created by similar encounters of such imperceptible bodies; especially when it seemed that no other way of explaining these things could be devised. And, to this end, things made by human skill helped me not a little: for I know of no distinction between these things and natural bodies, except that the operations of things made by skill are, for the most part, performed by apparatus large enough to be easily perceived by the senses: for this is necessary so that they can be made by men. On the other hand, however, natural effects almost always depend on some devices so minute that they escape all senses. And there are absolutely no judgments {or rules} in Mechanics which do not also pertain to Physics, of which Mechanics is a part or type: and it is as natural for a clock, composed of wheels of a certain kind, to indicate the hours, as for a tree, grown from a certain kind of seed, to produce the corresponding fruit. Accordingly, just as when those who are accustomed to considering automata know the use of some machine and see some of its parts, they easily conjecture from this how the other parts which they do not see are made: so, from the perceptible effects and parts of natural bodies, I have attempted to investigate the nature of their causes and of their imperceptible parts.

204.   That it suffices if I have explained what imperceptible things may be like, even if perhaps they are not so.

And although perhaps in this way it may be understood how all natural things could have been created, it should not therefore be concluded that they were in fact so created. For just as the same artisan can make two clocks which indicate the hours equally well and are exactly similar externally, but are internally composed of an entirely dissimilar combination of small wheels: so there is no doubt that the greatest Artificer of things could have made all those things which we see in many diverse ways. And indeed I most willingly concede this to be true, and will think that I have achieved enough if those things which I have written are only such that they correspond accurately to all the phenomena of nature, {whether these effects are produced by the causes I have explained or by others}. And indeed this will also suffice for the needs of everyday life, because Medicine and Mechanics, and all the other arts which can be perfected with the help of Physics, have as their goal only those effects which are perceptible and which accordingly ought to be numbered among the phenomena of nature. {And if these [desired] phenomena are produced by considering the consequences of some causes thus imagined, although false; we shall do as well as if these were the true causes, since the result is assumed similar as far as the perceptible effects are concerned}. And lest by chance anyone should believe that Aristotle ever achieved, or sought to achieve, anything more; he himself in the first book of the *Meteorology* at the beginning of Chapter 7, clearly asserts, concerning things which are not evident to the senses, that he thinks he is giving sufficient reasons and demonstrations if he only shows that these can be created as they are explained by him.

205.   That those things which I have explained here do seem at least morally certain, however.

However, lest some injury to truth may occur here, it must be considered that there are things which are held to be morally certain, that is, [certain] to a degree which suffices for the needs of everyday life; although if compared to the absolute power of God, they are uncertain. Thus, for example, if someone wishes to read a message written in Latin letters, to which however their true meaning has not been given and if, upon conjecturing that wherever there is an A in the message, a B must be read, and a C wherever there is a B, and that for each letter, the following one must be substituted; he finds that by this means certain Latin words are formed by these letters: he will not doubt that the true meaning of that message is contained in these words, even if he knows this solely by conjecture, and even though it may perhaps be the case that the person who wrote the message did not put the immediately following letters but some others in the place of the true ones, and thus concealed a different meaning in the message. It would however be so difficult for this to happen, {especially if the message contains many words}, that it does not seem credible. But those who notice how many things concerning the magnet, fire, and the fabric of the entire World have been deduced here from so few principles (even though they may suppose that I adopted these principles only by chance and without reason), will perhaps still know that it could scarcely have occurred that so many things should be consistent with one another, if they were false.

206.   That on the contrary they seem more than morally certain.

Besides, there are, even among natural things, some which we judge to be absolutely and more than morally certain; basing our judgment on the Metaphysical foundation that God is supremely good and by no means deceitful, and that, accordingly, the faculty which He gave us to distinguish the true from the false cannot err when we use it correctly and perceive something clearly with its help. Such are Mathematical demonstrations; such is the knowledge that material things exist; and such are all evident demonstrations which are made concerning material things. These reasonings of ours will perhaps be included among the number of these absolutely certain things by those who consider how they have been deduced in a continuous series from the first and simplest principles of human knowledge. Especially if they sufficiently understand that we can feel no external objects unless some local movement is excited by them in our nerves; and that such movement cannot be excited by the fixed stars, very far distant from here, unless some movement also occurs in these and in the whole intermediate heaven: for once these things have been accepted, it will scarcely seem possible for all the rest, at least the more general things which I have written about the World and the Earth, to be understood otherwise than as I have explained them.

207.   But that I submit all my opinions to the authority of the Church.

Nevertheless, mindful of my insignificance, I affirm nothing: but submit all these things both to the authority of the Catholic Church and to the judgment of men wiser than I; nor would I wish anyone to believe anything except what he is convinced of by clear and irrefutable reason.

Plate 1

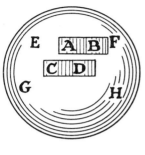

fig i

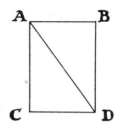

fig ii

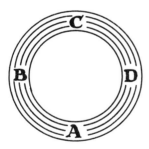

fig iii

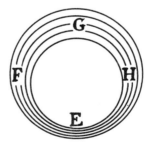

fig iv

Robiette delineavit

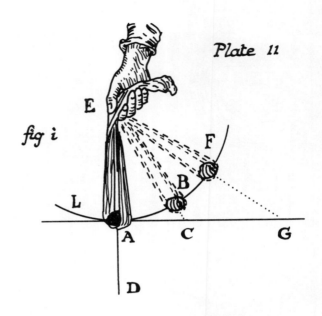

Plate 11

fig i

E

F

L

B

A   C   G

D

fig iii

fig ii

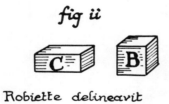

Robiette delineavit

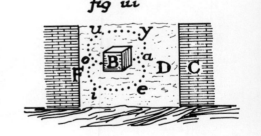

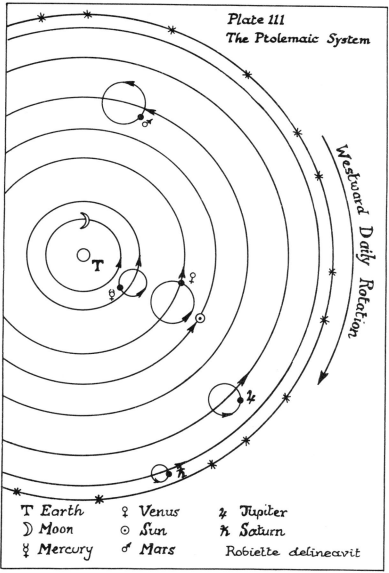

Plate III
The Ptolemaic System

Westward Daily Rotation

T Earth	♀ Venus	♃ Jupiter
☽ Moon	☉ Sun	♄ Saturn
☿ Mercury	♂ Mars	Robiette delineavit

The Earth is motionless at the center of the stellar sphere, and the entire system revolves westward once a day. The Moon, Sun, and planets also revolve eastward at various much slower rates. Each planet is placed on a small circle, its epicycle, which revolves at such a rate as to produce a periodic backward motion or retrogression of the planet. A variety of other devices was used to account for further irregularities in speed and position.

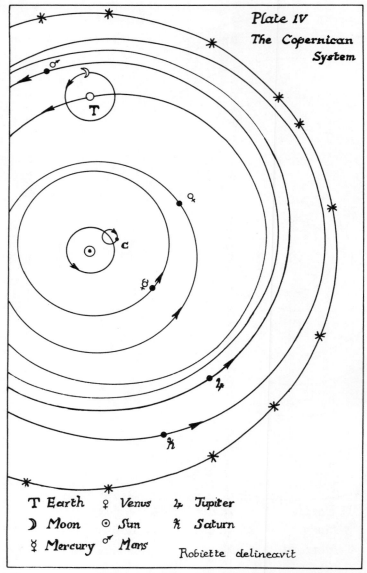

Plate IV
The Copernican System

T	Earth	♀	Venus	♃	Jupiter
☽	Moon	☉	Sun	♄	Saturn
☿	Mercury	♂	Mars		

Robiette delineavit

The Sun is at the center of the stellar sphere. The Earth rotates once a day from west to east and revolves eastward in a yearly circular orbit with its center at c. c itself moves on a deferent and epicycle system whose center is the Sun. The planets move eastward in circular orbits, the centers of which each move in such a way as to maintain a fixed relationship to c.

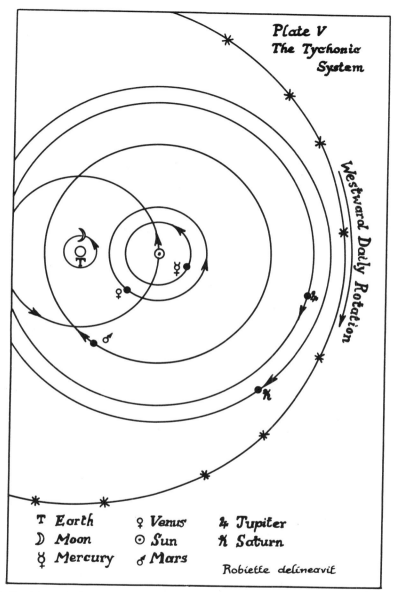

Plate V
The Tychonic
System

Westward Daily Rotation

♈ Earth	♀ Venus	♃ Jupiter
☽ Moon	☉ Sun	♄ Saturn
☿ Mercury	♂ Mars	

Robiette delineavit

The Earth is motionless at the center of the stellar sphere, and the stars, Moon, and Sun move as they do in the Ptolemaic system. The planets, however, move in orbits centered on the moving Sun; Mercury and Venus revolve eastward around the Sun, and the remaining planets revolve westward.

Plate VI

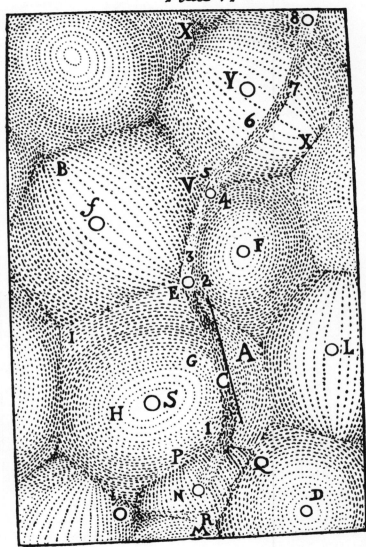

Plate VII

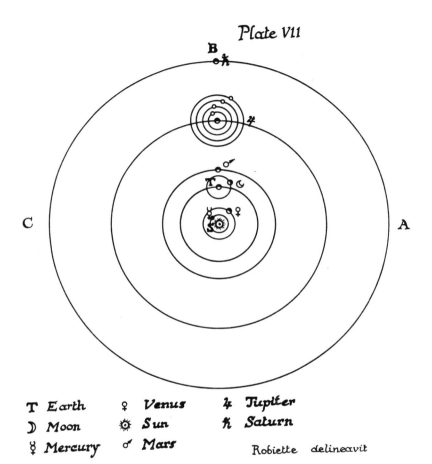

T Earth	♀ Venus	♃ Jupiter
☽ Moon	☉ Sun	♄ Saturn
☿ Mercury	♂ Mars	

Robiette delineavit

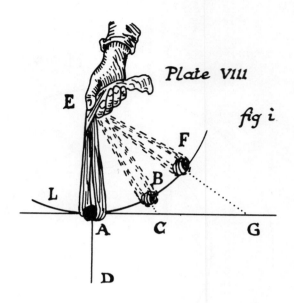

Plate VIII

fig i

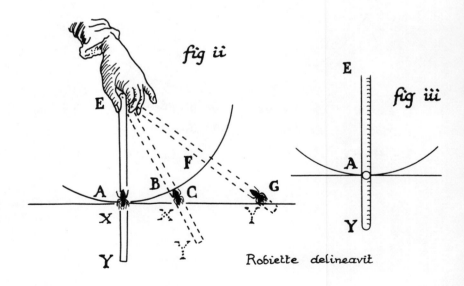

fig ii

fig iii

Robiette delineavit

Plate VIIII

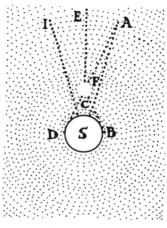

fig i

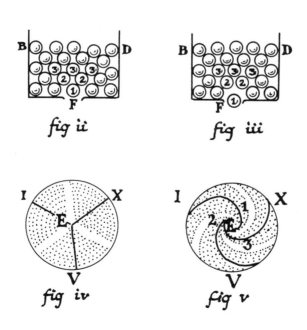

fig ii   fig iii

fig iv   fig v

Plate X

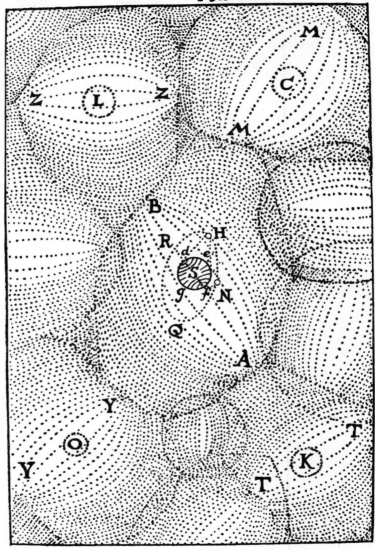

Plate X1

fig i

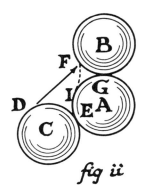

fig ii

fig iii

Robiette delineavit

Plate XII

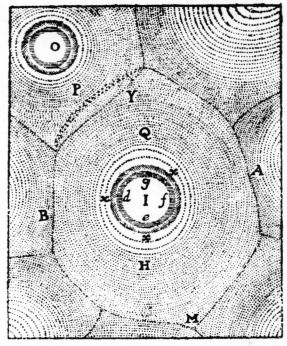

fig i

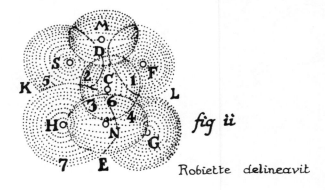

fig ii

Robiette delineavit

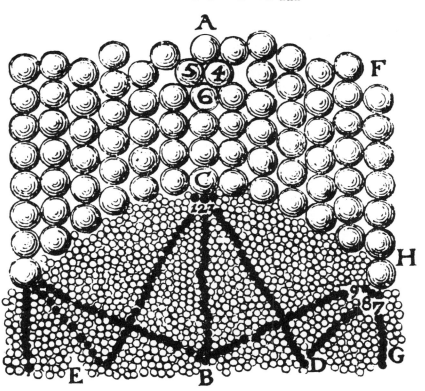

Plate XIII

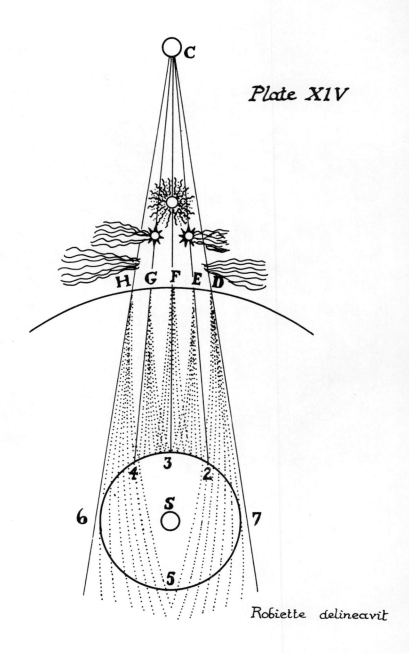

Plate XIV

Robiette delineavit

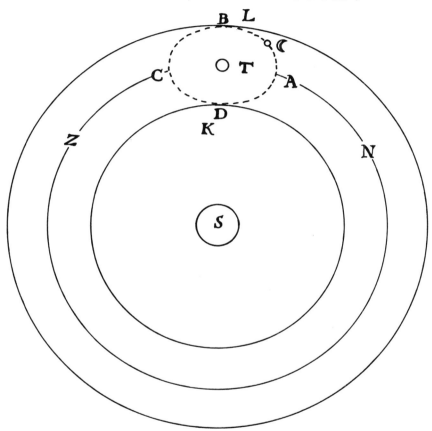

Plate XV

Robiette delineavit

Plate XVI

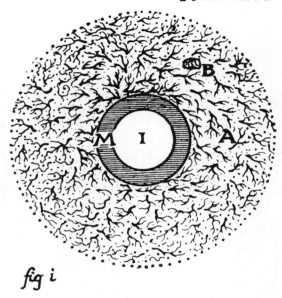

fig i

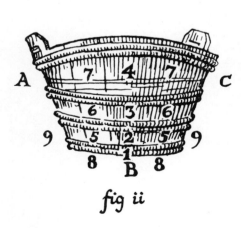

fig ii

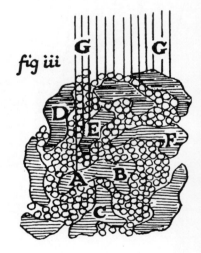

fig iii

Plate XVII

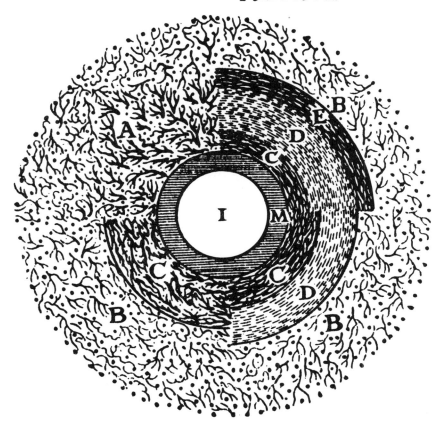

Plate XVIII

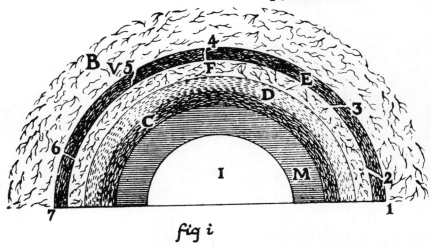

fig i

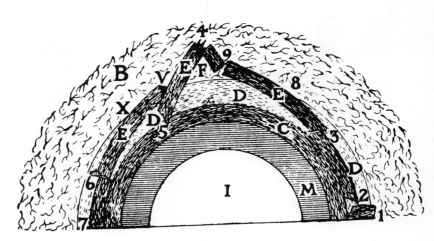

fig ii

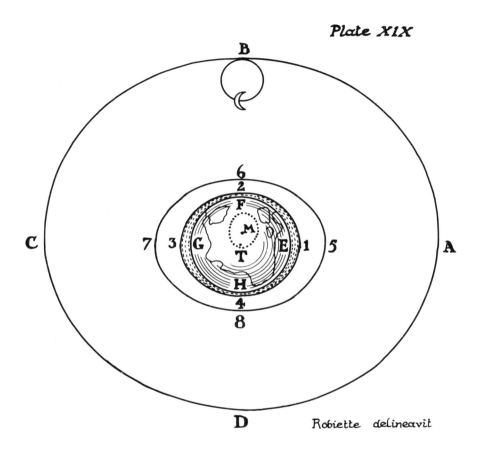

Plate XIX

Robiette delineavit

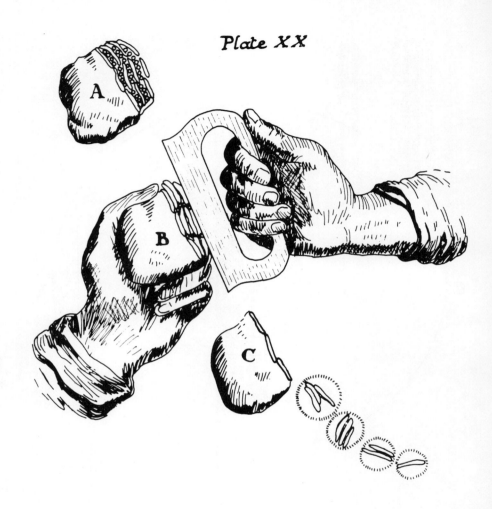

Plate XX

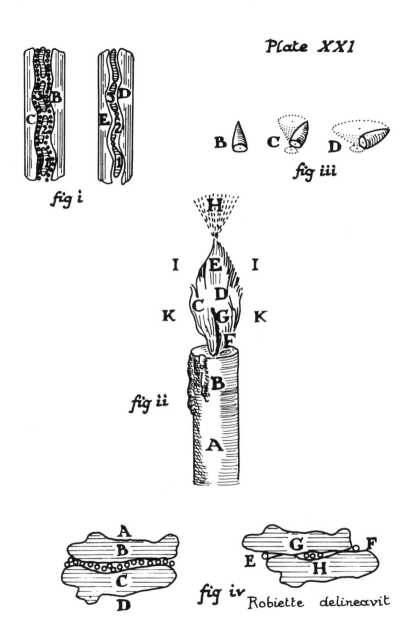

Plate XXI

fig i

fig iii

fig ii

fig iv   Robiette delineavit

Plate XXII

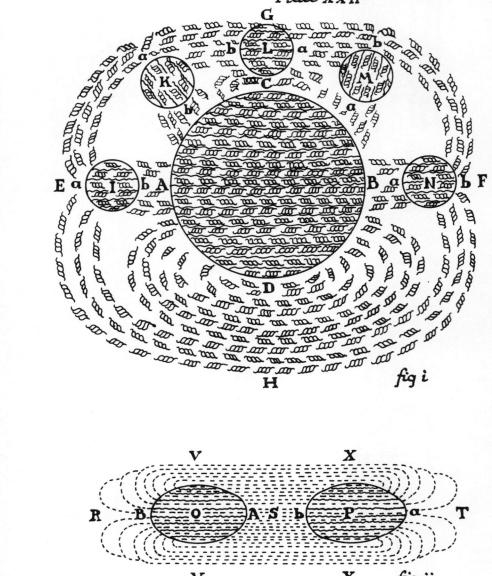

fig i

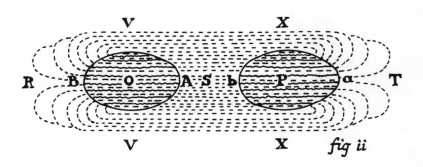

fig ii

Plate XXIII

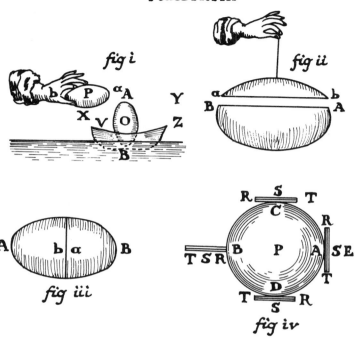

fig i

fig ii

fig iii

fig iv

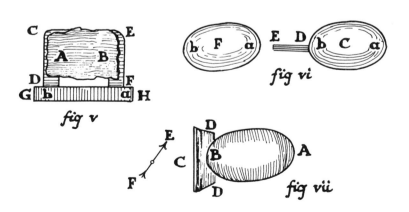

fig v

fig vi

fig vii

INDEX OF PERSONS AND WORKS

INDEX OF SUBJECTS AND TERMS

314

SYNTHESE HISTORICAL LIBRARY

Texts and Studies in the History of Logic and Philosophy

Editors:

N. KRETZMANN (Cornell University)
G. NUCHELMANS (University of Leyden)
L. M. DE RIJK (University of Leyden)

1. M. T. Beonio-Brocchieri Fumagalli, *The Logic of Abelard* (transl. from the Italian). 1969.
2. Gottfried Wilhelm Leibniz, *Philosophical Papers and Letters*. A selection translated and edited, with an introduction, by Leroy E. Loemker. 1969.
3. Ernst Mally, *Logische Schriften* (ed. by Karl Wolf and Paul Weingartner). 1971.
4. Lewis White Beck (ed.), *Proceedings of the Third International Kant Congress*. 1972.
5. Bernard Bolzano, *Theory of Science* (ed. by Jan Berg). 1973.
6. J. M. E. Moravcsik (ed.), *Patterns in Plato's Thought*. 1973.
7. Nabil Shehaby, *The Propositional Logic of Avicenna: A Translation from al-Shifa: al-Qiyas*, with Introduction, Commentary and Glossary. 1973.
8. Desmond Paul Henry, *Commentary on De Grammatico: The Historical-Logical Dimensions of a Dialogue of St. Anselm's*. 1974.
9. John Corcoran, *Ancient Logic and Its Modern Interpretations*. 1974.
10. E. M. Barth, *The Logic of the Articles in Traditional Philosophy*. 1974.
11. Jaakko Hintikka, *Knowledge and the Known. Historical Perspectives in Epistemology*. 1974.
12. E. J. Ashworth, *Language and Logic in the Post-Medieval Period*. 1974.
13. Aristotle, *The Nicomachean Ethics* (transl. with Commentaries and Glossary by Hypocrates G. Apostle). 1975.
14. R. M. Dancy, *Sense and Contradiction: A Study in Aristotle*. 1975.
15. Wilbur Richard Knorr, *The Evolution of the Euclidean Elements. A Study of the Theory of Incommensurable Magnitudes and Its Significance for Early Greek Geometry*. 1975.
16. Augustine, *De Dialectica* (transl. with Introduction and Notes by B. Darrell Jackson). 1975.
17. Arpád Szabó, *The Beginnings of Greek Mathematics*. 1978.
18. Rita Guerlac, *Juan Luis Vives Against the Pseudodialecticians. A Humanist Attack on Medieval Logic*. Texts, with translation, introduction and notes. 1979.
19. Paul Vincent Spade, *Peter of Ailly: Concepts and Insolubles. An Annotated Translation*. 1980.
20. Simo Knuuttila (ed.), *Reforging the Great Chain of Being*. 1981.
21. Jill Vance Buroker, *Space and Incongruence*. 1981.
22. E. P. Bos, *Marsilius of Inghen*. 1982.
23. William Remmelt de Jong, *The Semantics of John Stuart Mill*. 1982